ENCYCLOPAEDIA OF BIOTECHNOLOGY & GENETIC ENGINEERING

ENCYCLOPAEDIA OF
BIOTECHNOLOGY & GENETIC ENGINEERING

Bipin Kumar Panda

ANMOL PUBLICATIONS PVT. LTD.
NEW DELHI - 110 002 (INDIA)

ANMOL PUBLICATIONS PVT. LTD.
H.O.: 4374/4B, Ansari Road, Daryaganj,
New Delhi-110 002 (India)
Ph.: 23278000, 23261597
B.O.: No. 1015, Ist Main Road, BSK IIIrd Stage
IIIrd Phase, IIIrd Block,
Bangalore - 560 085 (India)
Visit us at: www.anmolpublications.com

Encyclopaedia of Biotechnology & Genetic Engineering

ISBN 978-81-261-3593-6

PRINTED IN INDIA

Printed at Mehra Offset Press, Delhi.

Contents

Preface

This publication titled "Encyclopaedia of Biotechnology & Genetic Engineering" is an initiative to deal with the subject in its totality. Biotech refers to the biological science, applied especially in genetic engineering and recombinant DNA technology. Biotechnology deals with a set of biological techniques developed through basic research and now applied to research and product development. It is the means or way of manipulating life forms (organisms) to provide desirable products for man's use. For example, beekeeping and cattle breeding could be considered to be biotechnology-related endeavors. However, usage of the word biotechnology in the United States has come to mean all parts of an industry that knowingly create, develop, and market a variety of products through the willful manipulation, on a molecular level, of life forms or utilization of knowledge pertaining to living systems. A common misconception is that biotechnology refers only to recombinant DNA (rDNA) work. However, recombinant DNA is only one of the many techniques used to derive products from organisms, plants, and parts of both for the biotechnology industry. A list of areas covered by the term biotechnology would more properly include: recombinant DNA, plant tissue culture, rDNA or gene splicing, enzyme systems, plant breeding, meristem culture, mammalian cell culture, immunology, molecular biology, fermentation, and others. Today, all these are a rapidly growing field of research and development that is cutting across many traditional boundaries. Genetic engineering, recombinant DNA technology, genetic modification/manipulation (GM) and gene splicing are terms that are applied to the direct manipulation of an organism's genes. Genetic Engineering is not to be confused with traditional breeding where the organisims genes are manipulated *indirectly*. Genetic engineering uses the techniques of molecular cloning and transformation. Genetic Engineering endeavors have found success in improving crop technology, the manufacture of human insulin through the use of modified bacteria, the manufacture of erythropoietin in Chinese hamster ovary cells, and the production of new types of experimental mice such as the oncomouse (cancer mouse) for research. Since a protein sequence is specified by a segment of DNA called a gene, novel versions of that protein can be produced by changing the DNA sequence of the gene. Some groups have argued that genetic engineering is wrong and is "doing the work of God", but most scientists believe that genetic engineering is essential to help future medical discoveries. However, even with regard to this technology's great potential, some people have raised concerns about the introduction of genetically engineered plants and animals into the environment and the potential dangers of human consumption of GM foods. They say that these organisms have the potential to spread their modified genes into native populations thereby disrupting natural ecosystems.

This encyclopaedia titled "Encyclopaedia of Biotechnology & Genetic Engineering" is really unique in terms of coverage, scope, elaboration, usefulness and futuristic perspective. It will provide readers with a user-friendly approach towards acquainting themselves with various select traditional and newly emerging fields of biosciences. This encyclopaedia is of great contemporary relevance. It will serve as a reference book on the said subject and tends to acquaint readers with both in-depth and extensive knowledge of the subject concerned.

This encyclopaedia is a unique initiative to provide the most comprehensive coverage to the subject area in question. This encyclopedia mainly focuses on the following areas: Biotechnology and Its Historical Timeline; Biotechnology and Genetic Resource: Uses, Research Areas and Regulations; Gene Science and Science of Genetics; Gene Related Processes, Genetic Changes and Cloning; Overview of Genetic Engineering & Bioengineering; GMOs With Special Focus on Genetically Modified Food etc.

—Editor

1

Introduction to Biotechnology and Its Historical Timeline

BIOTECHNOLOGY

Biotechnology is technology based on biology, especially when used in agriculture, food science, and medicine. The United Nations Convention on Biological Diversity has come up with one of many definitions of biotechnology:

> *"Biotechnology has contributed towards the exploitation of biological organisms or biological processes through modern techniques, which could be profitably used in medicine, agriculture, animal husbandry and environmental cloning."*

Biotechnology is a popular term for the generic technology of the 21st century. Although it has been utilized for centuries in traditional production processes, modern biotechnology is only 50 years old and in the last decades it has been witnessing tremendous developments. Bioengineering is the science upon which all Biotechnological applications are based. With the development of new approaches and modern techniques, traditional biotechnology industries are also acquiring new horizons enabling them to improve the quality of their products and increase the productivity of their systems.

Before 1971, the term, *biotechnology*, was primarily used in the food processing and agriculture industries. Since the 1970s, it began to be used by the Western scientific establishment to refer to laboratory-based techniques being developed in biological research, such as recombinant DNA or tissue culture-based processes. In fact, the term should be used in a much broader sense to describe the whole range of methods, both ancient and modern, used to manipulate organic materials to reach the demands of food production. So the term could be defined as, "The application of indigenous and/or scientific knowledge to the management of (parts of) microorganisms, or of cells and tissues of higher organisms, so that these supply goods and services of use to the food industry and its consumers.

There has been a great deal of talk—and money—poured into biotechnology with the hope that miracle drugs will appear. While there do seem to be a small number of efficacious drugs, in general the biotech revolution has not happened in the pharmaceutical sector. However, recent progress with monoclonal antibody based drugs, such as Genentech's Avastin suggest that biotech may finally have found a role in pharmaceutical sales.

Biotechnology combines disciplines like genetics, molecular biology, biochemistry, embryology and cell biology, which are in turn linked to practical disciplines like chemical engineering, information technology, and robotics.

History

The most practical use of biotechnology, which is still present today, is the cultivation of plants to produce food suitable to humans. Agriculture has been theorized to have become the dominant way of producing food since the Neolithic Revolution. The processes and methods of agriculture have been refined by other mechanical and biological sciences since its inception. Through early biotechnology farmers were able to select the best suited and highest-yield crops to produce enough food to support a growing population. Other uses of biotechnology were required as crops and fields became increasingly large and difficult to maintain. Specific organisms and organism byproducts were used to fertilize, restore nitrogen, and control pests. Throughout the use of agriculture farmers have inadvertently altered the genetics of their crops through introducing them to new environments and breeding them with other plants—one of the first forms of biotechnology. Cultures such as those in Mesopotamia, Egypt, and Iran developed the process of brewing beer. It is still done by the same basic method of using malted grains (containing enzymes) to convert starch from grains into sugar and then adding specific yeasts to produce beer. In this process the carbohydrates in the grains were broken down into alcohols such as ethanol. Later other cultures produced the process of Lactic acid fermentation which allowed the fermentation and preservation of other forms of food. Fermentation was also used in this time period to produce leavened bread. Although the process of fermentation was not fully understood until Louis Pasteur's work in 1857, it is still the first use of biotechnology to convert a food source into another form.

Combinations of plants and other organisms were used as medications in many early civilizations. Since as early as 200 BC, people began to use disabled or minute amounts of infectious agents to immunize themselves against infections. These and similar processes have been refined in modern medicine and have lead to many developments such as antibiotics, vaccines, and other methods of fighting sickness.

In the early twentieth century scientists gained a greater understanding of microbiology and explored ways of manufacturing specific products. In 1917, Chaim Weizmann first used a pure microbiological culture in an industrial process, that of manufacturing corn starch using *Clostridium acetobutylicum* to produce acetone, which the United Kingdom desperately needed to manufacture explosives during World War I.

The field of modern biotechnology is thought to have largely begun on June 16, 1980, when the United States Supreme Court ruled that a genetically-modified microorganism could be patented in the case of *Diamond v. Chakrabarty*. Indian-born Ananda Chakrabarty, working for General Electric, had developed a bacterium (derived from the *Pseudomonas* genus) capable of breaking down crude oil, which he proposed to use in treating oil spills. A university in Florida is now studying ways to prevent tooth decay. They altered the bacteria in the tooth called *Streptococcus mutans* by stripping it down so it could not produce lactic acid.

Applications

Biotechnology has applications in four major industrial areas, including health care, crop production and agriculture, non food uses of crops (e.g. biodegradable plastics, vegetable oil, biofuels), and environmental uses. For example, one application of biotechnology is the directed use of organisms for the manufacture of organic products (examples include beer and milk products). Another example is using naturally present bacteria by the mining industry in bioleaching. Biotechnology is also used to recycle, treat waste, clean up sites contaminated by industrial activities (bioremediation), and also to produce biological weapons.

Red biotechnology is applied to medical processes. Some examples are the designing of organisms to produce antibiotics, and the engineering of genetic cures through genomic manipulation.

White biotechnology , also known as industrial biotechnology, is biotechnology applied to industrial processes. An example is the designing of an organism to produce a useful chemical. Another example is the using of enzymes as industrial catalysts to either produce valuable chemicals or destroy hazardous/polluting chemicals (examples using oxidoreductases are given in Feng Xu (2005) "Applications of oxidoreductases: Recent progress" Ind. Biotechnol. 1, 38-50). White biotechnology tends to consume less in resources than traditional processes used to produce industrial goods.

Green biotechnology is biotechnology applied to agricultural processes. An example is the designing of transgenic plants to grow under specific environmental conditions or in the presence (or absence) of certain agricultural chemicals. One hope is that green biotechnology might produce more environmentally friendly solutions than traditional industrial agriculture. An example of this is the engineering of a plant to express a pesticide, thereby eliminating the need for external application of pesticides. An example of this would be Bt corn. Whether or not green biotechnology products such as this are ultimately more environmentally friendly is a topic of considerable debate.

The term blue biotechnology has also been used to describe the marine and aquatic applications of biotechnology, but its use is relatively rare.

The investments and economic output of all of these types of applied biotechnologies form what has been described as the bioeconomy.

Bioinformatics is an interdisciplinary field which addresses biological problems using computational techniques, and makes the rapid organization and analysis of biological data possible. The field may also be referred to as *computational biology*, and can be defined as, "conceptualizing biology in terms of molecules and then applying informatics techniques to understand and organize the information associated with these molecules, on a large scale." Bioinformatics plays a key role in various areas, such as functional genomics, structural genomics, and proteomics, and forms a key component in the biotechnology and pharmaceutical sector.

Medicine

In medicine, modern biotechnology finds promising applications in such areas as

- pharmacogenomics;
- drug production;
- genetic testing; and
- gene therapy.

Pharmacogenomics

Pharmacogenomics is the study of how the genetic inheritance of an individual affects his/her body's response to drugs. It is a coined word derived from the words "pharmacology" and "genomics". It is hence the study of the relationship between pharmaceuticals and genetics. The vision of pharmacogenomics is to be able to design and produce drugs that are adapted to each person's genetic makeup.

Pharmacogenomics results in the following benefits:

1. Development of tailor-made medicines. Using pharmacogenomics,

pharmaceutical companies can create drugs based on the proteins, enzymes and RNA molecules that are associated with specific genes and diseases. These tailor-made drugs promise not only to maximize therapeutic effects but also to decrease damage to nearby healthy cells.

2. More accurate methods of determining appropriate drug dosages. Knowing a patient's genetics will enable doctors to determine how well his/ her body can process and metabolize a medicine. This will maximize the value of the medicine and decrease the likelihood of overdose.Ã
3. Improvements in the drug discovery and approval process. The discovery of potential therapies will be made easier using genome targets. Genes have been associated with numerous diseases and disorders. With modern biotechnology, these genes can be used as targets for the development of effective new therapies, which could significantly shorten the drug discovery process.
4. Better vaccines. Safer vaccines can be designed and produced by organisms transformed by means of genetic engineering. These vaccines will elicit the immune response without the attendant risks of infection. They will be inexpensive, stable, easy to store, and capable of being engineered to carry several strains of pathogen at once.

Pharmaceutical Products

Traditional pharmaceutical drugs are relatively simple molecules that have been found primarily through trial and error to treat the symptoms of a disease or illness. Biopharmaceuticals are large biological molecules known as proteins and these target the underlying mechanisms and pathways of a malady; it is a relatively young industry. They can deal with targets in humans that are not accessible with traditional medicines. A patient typically is dosed with a small molecule *via* a tablet while a large molecule is typically injected.

Small molecules are manufactured by chemistry but large molecules are created by living cells: for example, bacteria cells, yeast cell, and animal cells.

Modern biotechnology is often associated with the use of genetically altered microorganisms such as *E. coli* or yeast for the production of substances like insulin or antibiotics. It can also refer to transgenic animals or transgenic plants, such as Bt corn. Genetically altered mammalian cells, such as Chinese Hamster Ovary (CHO) cells, are also widely used to manufacture pharmaceuticals. Another promising new biotechnology application is the development of plant-made pharmaceuticals.

Biotechnology is also commonly associated with landmark breakthroughs in new medical therapies to treat diabetes, hepatitis B, hepatitis C, cancers, arthritis, haemophilia, bone fractures, multiple sclerosis, cardiovascular as well as molecular diagnostic devices than can be used to define the patient population. Herceptin, is the first drug approved for use with a matching diagnostic test and is used to treat breast cancer in women whose cancer cells express the protein HER2.

Modern biotechnology can be used to manufacture existing drugs more easily and cheaply. The first genetically engineered products were medicines designed to combat human diseases. To cite one example, in 1978 Genentech joined a gene for insulin and a plasmid vector and put the resulting gene into a bacterium called *Escherichia coli*. Insulin, widely used for the treatment of diabetes, was previously extracted from the pancreas of bovines and pigs. It was very expensive and often elicited unwanted allergic responses. The resulting genetically engineered bacterium enabled the production of vast quantities of human insulin at low cost.

Since then modern biotechnology has made it possible to produce more easily and cheaply human growth hormone, clotting factors for hemophiliacs, fertility drugs, erythropoietin and other drugs. Most drugs today are based on about 500 molecular targets. Genomic knowledge of the genes involved in diseases, disease pathways, and drug-response sites are expected to lead to the discovery of thousands more new targets.

Genetic Testing

Genetic testing involves the direct examination of the DNA molecule itself. A scientist scans a patient's DNA sample for mutated sequences.

There are two major types of gene tests. In the first type, a researcher may design short pieces of DNA ("probes") whose sequences are complementary to the mutated sequences. These probes will seek their complement among the base pairs of an individual's genome. If the mutated sequence is present in the patient's genome, the probe will bind to it and flag the mutation. In the second type, a researcher may conduct the gene test by comparing the sequence of DNA bases in a patient's gene to disease in healthy individuals or their progeny.

Genetic testing is now used for:

- Determining sex
- Carrier screening, or the identification of unaffected individuals who carry one copy of a gene for a disease that requires two copies for the disease to manifest
- Prenatal diagnostic screening
- Newborn screening

- Presymptomatic testing for predicting adult-onset disorders
- Presymptomatic testing for estimating the risk of developing adult-onset cancers
- Confirmational diagnosis of symptomatic individuals
- Forensic/identity testing

Some genetic tests are already available, although most of them are used in developed countries. The tests currently available can detect mutations associated with rare genetic disorders like cystic fibrosis, sickle cell anemia, and Huntington's disease. Recently, tests have been developed to detect mutation for a handful of more complex conditions such as breast, ovarian, and colon cancers. However, gene tests may not detect every mutation associated with a particular condition because many are as yet undiscovered, and the ones they do detect may present different risks to different people and populations.

Gene Therapy

Gene therapy may be used for treating, or even curing, genetic and acquired diseases like cancer and AIDS by using normal genes to supplement or replace defective genes or to bolster a normal function such as immunity. It can be used to target somatic (i.e., body) or germ (i.e., egg and sperm) cells. In somatic gene therapy, the genome of the recipient is changed, but this change is not passed along to the next generation. In contrast, in germline gene therapy, the egg and sperm cells of the parents are changed for the purpose of passing on the changes to their offspring.

There are basically two ways of implementing a gene therapy treatment:

1. *Ex vivo*, which means "outside the body" – Cells from the patient's blood or bone marrow are removed and grown in the laboratory. They are then exposed to a virus carrying the desired gene. The virus enters the cells, and the desired gene becomes part of the DNA of the cells. The cells are allowed to grow in the laboratory before being returned to the patient by injection into a vein.
2. *In vivo*, which means "inside the body" – No cells are removed from the patient's body. Instead, vectors are used to deliver the desired gene to cells in the patient's body.

Currently, the use of gene therapy is limited. Somatic gene therapy is primarily at the experimental stage. Germline therapy is the subject of much discussion but it is not being actively investigated in larger animals and human beings.

As of June 2001, more than 500 clinical gene-therapy trials involving about 3,500 patients have been identified worldwide. Around 78% of these are in the United States, with Europe having 18%. These trials focus on various types of cancer, although other multigenic diseases are being studied as well. Recently, two children born with severe combined immunodeficiency disorder ("SCID") were reported to have been cured after being given genetically engineered cells.

Gene therapy faces many obstacles before it can become a practical approach for treating disease. At least four of these obstacles are as follows:

1. *Gene delivery tools*. Genes are inserted into the body using gene carriers called vectors. The most common vectors now are viruses, which have evolved a way of encapsulating and delivering their genes to human cells in a pathogenic manner. Scientists manipulate the genome of the virus by removing the disease-causing genes and inserting the therapeutic genes. However, while viruses are effective, they can introduce problems like toxicity, immune and inflammatory responses, and gene control and targeting issues.
2. *Limited knowledge of the functions of genes*. Scientists currently know the functions of only a few genes. Hence, gene therapy can address only some genes that cause a particular disease. Worse, it is not known exactly whether genes have more than one function, which creates uncertainty as to whether replacing such genes is indeed desirable.
3. *Multigene disorders and effect of environment*. Most genetic disorders involve more than one gene. Moreover, most diseases involve the interaction of several genes and the environment. For example, many people with cancer not only inherit the disease gene for the disorder, but may have also failed to inherit specific tumor suppressor genes. Diet, exercise, smoking and other environmental factors may have also contributed to their disease.
4. *High costs*. Since gene therapy is relatively new and at an experimental stage, it is an expensive treatment to undertake. This explains why current studies are focused on illnesses commonly found in developed countries, where more people can afford to pay for treatment. It may take decades before developing countries can take advantage of this technology.

Human Genome Project

The Human Genome Project is an initiative of the U.S. Department of Energy ("DOE") that aims to generate a high-quality reference sequence for the entire human genome and identify all the human genes.

The DOE and its predecessor agencies were assigned by the U.S. Congress to develop new energy resources and technologies and to pursue a deeper understanding of potential health and environmental risks posed by their production and use. In 1986, the DOE announced its Human Genome Initiative. Shortly thereafter, the DOE and National Institutes of Health developed a plan for a joint Human Genome Project ("HGP"), which officially began in 1990.

The HGP was originally planned to last 15 years. However, rapid technological advances and worldwide participation accelerated the completion date to 2005. Already it has enabled gene hunters to pinpoint genes associated with more than 30 disorders.

Cloning

Human cloning is one of the techniques of modern biotechnology. It involves the removal of the nucleus from one cell and its placement in an unfertilized egg cell whose nucleus has either been deactivated or removed.

There are two types of cloning:

1. Reproductive cloning. After a few divisions, the egg cell is placed into a uterus where it is allowed to develop into a fetus that is genetically identical to the donor of the original nucleus.
2. Therapeutic cloning. The egg is placed into a Petri dish where it develops into embryonic stem cells, which have shown potentials for treating several ailments.

The major differences between these two types are shown in Table 1.

In February 1997, cloning became the focus of media attention when Ian Wilmut and his colleagues at the Roslin Institute announced the successful cloning of a sheep, named Dolly, from the mammary glands of an adult female. The cloning of Dolly made it apparent to many that the techniques used to produce her could someday be used to clone human beings. This stirred a lot of controversy because of its ethical implications.

Concerns Regarding the use of Modern Biotechnology Techniques in Medicine

Several issues have been raised regarding the use of modern biotechnology in the medical sector. Many of these issues are similar to those facing any new technology that is viewed as powerful and far-reaching. Some of these issues are:

1. Absence of cure. There is still a lack of effective treatment or preventive measures for many diseases and conditions now being diagnosed or predicted using gene tests. Thus, revealing information about risk of a future

disease that has no existing cure presents an ethical dilemma for medical practitioners.

2. Ownership and control of genetic information. Who will own and control genetic information, or information about genes, gene products, or inherited characteristics derived from an individual or a group of people like indigenous communities? At the macro level, there is a possibility of a genetic divide, with developing countries that do not have access to medical applications of biotechnology being deprived of benefits accruing from products derived from genes obtained from their own people. Moreover, genetic information can pose a risk for minority population groups as it can lead to group stigmatization.

   At the individual level, the absence of privacy and anti-discrimination legal protections in most countries can lead to discrimination in employment or insurance or other misuse of personal genetic information. This raises questions such as whether genetic privacy is different from medical privacy.

3. Reproductive issues. These include the use of genetic information in reproductive decision-making and the possibility of genetically altering reproductive cells that may be passed on to future generations. For example, germline therapy forever changes the genetic make-up of an individual's descendants. Thus, any error in technology or judgment may have far-reaching consequences. Ethical issues like designer babies and human cloning have also given rise to controversies between and among scientists and bioethicists, especially in the light of past abuses with eugenics.

4. Clinical issues. These center on the capabilities and limitations of doctors and other health-service providers, people identified with genetic conditions, and the general public in dealing with genetic information. For instance, how should the public be prepared to make informed choices based on the results of genetic tests? How will genetic tests be evaluated and regulated for accuracy, reliability, and usefulness?

5. Effects on social institutions. Genetic tests reveal information about individuals and their families. Thus, test results can affect the dynamics within social institutions, particularly the family.

6. Conceptual and philosophical implications regarding human responsibility, free will vis-à-vis genetic determinism, and the concepts of health and disease. Do genes influence human behavior? If so, does genetic testing mean controlling human behavior? What is considered acceptable diversity? What is normal and what is a disability or disorder, and who decides these matters? Are disabilities diseases that need to be cured or prevented?

Where should the line between medical treatment and enhancement be drawn; moreover, what considerations make that delineation important and how do such considerations bear on public policy and personal choice? Who will have access to gene therapy?

Agriculture

There are many applications of biotechnology in agriculture.

One is improved yield from crops. Using the techniques of modern biotechnology, one or two genes may be transferred to a highly developed crop variety to impart a new character that would increase its yield. However, while increases in crop yield are the most obvious applications of modern biotechnology in agriculture, it is also the most difficult one. Current genetic engineering techniques work best for effects that are controlled by a single gene. Many of the genetic characteristics associated with yield (e.g., enhanced growth) are controlled by a large number of genes, each of which has a minimal effect on the overall yield. There is, therefore, much scientific work to be done in this area.

Another is the reduced vulnerability of crops to environmental stresses. Crops containing genes that will enable them to withstand biotic and abiotic stresses may be developed. For example, drought and excessively salty soil are two important limiting factors in crop productivity. Biotechnologists are studying plants that can cope with these extreme conditions in the hope of finding the genes that enable them to do so and eventually transferring these genes to the more desirable crops. One of the latest developments is the identification of a plant gene, At-DBF2, from thale cress, a tiny weed that is often used for plant research because it is very easy to grow and its genetic code is well mapped out. When this gene was inserted into tomato and tobacco cells, the cells were able to withstand environmental stresses like salt, drought, cold and heat, far more than ordinary cells. If these preliminary results prove successful in larger trials, then At-DBF2 genes can help in engineering crops that can better withstand harsh environments.

Researchers have also created transgenic rice plants that are resistant to rice yellow mottle virus (RYMV). In Africa, this virus destroys majority of the rice crops and makes the surviving plants more susceptible to fungal infections.

Increased nutritional qualities of food crops. Proteins in foods may be modified to increase their nutritional qualities. Proteins in legumes and cereals may be transformed to provide the amino acids needed by human beings for a balanced diet. A good example is the work of Professors Ingo Potrykus and Peter Beyer on the so-called Goldenrice.

Improved taste, texture or appearance of food. Modern biotechnology can be used to slow down the process of spoilage so that fruit can ripen longer on the plant and then be transported to the consumer with a still reasonable shelf life. This improves the taste, texture and appearance of the fruit. More importantly, it could expand the market for farmers in developing countries due to the reduction in spoilage.

The first genetically modified food product was a tomato which was transformed to delay its ripening. Researchers in Indonesia, Malaysia, Thailand, Philippines and Vietnam are currently working on delayed-ripening papaya in collaboration with the University of Nottingham and Zeneca.

Reduced dependence on fertilizers, pesticides and other agrochemicals. Most of the current commercial applications of modern biotechnology in agriculture are on reducing the dependence of farmers on agrochemicals. For example, *Bacillus thuringiensis* (Bt) is a soil bacterium that produces a protein with insecticidal qualities. Traditionally, a fermentation process has been used to produce an insecticidal spray from these bacteria. In this form, the Bt toxin occurs as an inactive protoxin, which requires digestion by an insect to be effective. There are several Bt toxins and each one is specific to certain target insects. Crop plants have now been engineered to contain and express the genes for Bt toxin, which they produce in its active form. When a susceptible insect ingests the transgenic crop cultivar expressing the Bt protein, it stops feeding and soon thereafter dies as a result of the Bt toxin binding to its gut wall. Bt corn is now commercially available in a number of countries to control corn borer (a lepidopteran insect), which is otherwise controlled by spraying

Crops have also been genetically engineered to acquire tolerance to broad-spectrum herbicide. The lack of cost-effective herbicides with broad-spectrum activity and no crop injury was a consistent limitation in crop weed management. Multiple applications of numerous herbicides were routinely used to control a wide range of weed species detrimental to agronomic crops. Weed management tended to rely on preemergence — that is, herbicide applications were sprayed in response to expected weed infestations rather than in response to actual weeds present. Mechanical cultivation and hand weeding were often necessary to control weeds not controlled by herbicide applications. The introduction of herbicide tolerant crops has the potential of reducing the number of herbicide active ingredients used for weed management, reducing the number of herbicide applications made during a season, and increasing yield due to improved weed management and less crop injury. Transgenic crops that express tolerance to glyphosphate, glufosinate and bromoxynil have been developed. These herbicides can now be sprayed on transgenic crops without inflicting damage on the crops while killing nearby weeds.

From 1996 to 2001, herbicide tolerance was the most dominant trait introduced to commercially available transgenic crops, followed by insect resistance. In 2001, herbicide tolerance deployed in soybean, corn and cotton accounted for 77% of the 626,000 square kilometres planted to transgenic crops; Bt crops accounted for 15%; and "stacked genes" for herbicide tolerance and insect resistance used in both cotton and corn accounted for 8%.

Production of novel substances in crop plants. Modern biotechnology is increasingly being applied for novel uses other than food. For example, oilseed is at present used mainly for margarine and other food oils, but it can be modified to produce fatty acids for detergents, substitute fuels and petrochemicals. Banana trees and tomato plants have also been genetically engineered to produce vaccines in their fruit. If future clinical trials prove successful, the advantages of edible vaccines would be enormous, especially for developing countries. The transgenic plants may be grown locally and cheaply. Homegrown vaccines would also avoid logistical and economic problems posed by having to transport traditional preparations over long distances and keeping them cold while in transit. And since they are edible, they will not need syringes, which are not only an additional expense in the traditional vaccine preparations but also a source of infections if contaminated.

There is another, darker side, many people say, to the agricultural biotechnology issue however. It includes increased herbicide usage and resultant herbicide resistance, "super weeds," residues on and in food crops, genetic contamination of non-GM crops which hurt organic and conventional farmers, damage to wildlife from glyphosate, mass suicides in India, corruption, rampant monopolism etc.

Biological Engineering

Biotechnological engineering or biological engineering is a branch of engineering that focuses on biotechnologies and biological science. It includes different disciplines such as biochemical engineering, biomedical engineering, bio-process engineering, biosystem engineering and so on. Because of the novelty of the field, the definition of a bioengineer is still undefined. However, in general it is an integrated approach of fundamental biological sciences and traditional engineering principles.

Bioengineers are often employed to scale up bio processes from the laboratory scale to the manufacturing scale. Moreover, as with most engineers, they often deal with management, economic and legal issues. Since patents and regulation (e.g. FDA regulation in the U.S.) are very important issues for biotech enterprises, bioengineers are often required to have knowledge related to these issues.

The increasing number of biotech enterprises is likely to create a need for bioengineers in the years to come. Many universities throughout the world are now providing programs in bioengineering and biotechnology (as independent programs or specialty programs within more established engineering fields).

Bioremediation and Biodegradation

Biotechnology is being used to engineer and adapt organisms especially microorganisms in an effort to find sustainable ways to clean up contaminated environments. The elimination of a wide range of pollutants and wastes from the environment is an absolute requirement to promote a sustainable development of our society with low environmental impact. Biological processes play a major role in the removal of contaminants and biotechnology is taking advantage of the astonishing catabolic versatility of microorganisms to degrade/convert such compounds. New methodological breakthroughs in sequencing, genomics, proteomics, bioinformatics and imaging are producing vast amounts of information. In the field of Environmental Microbiology, genome-based global studies open a new era providing unprecedented *in silico* views of metabolic and regulatory networks, as well as clues to the evolution of degradation pathways and to the molecular adaptation strategies to changing environmental conditions. Functional genomic and metagenomic approaches are increasing our understanding of the relative importance of different pathways and regulatory networks to carbon flux in particular environments and for particular compounds and they will certainly accelerate the development of bioremediation technologies and biotransformation processes.

Marine environments are especially vulnerable since oil spills of coastal regions and the open sea are poorly containable and mitigation is difficult. In addition to pollution through human activities, millions of tons of petroleum enter the marine environment every year from natural seepages. Despite its toxicity, a considerable fraction of petroleum oil entering marine systems is eliminated by the hydrocarbon-degrading activities of microbial communities, in particular by a remarkable recently discovered group of specialists, the so-called hydrocarbonoclastic bacteria (HCB).

HISTORY OF BIOTECHNOLOGY

Biotechnology is the application of scientific and engineering principles to the processing of materials by biological agents to provide goods and services. From its inception, biotechnology has maintained a close relationship with society. Although now most often associated with the development of remarkable drugs, historically biotechnology has been principally associated with food, addressing such issues as malnutrition and famine. The history of biotechnology begins with zymotechnology, which commenced with a focus on brewing techniques for beer. By World War I, however, zymotechnology would expand to tackle larger industrial issues, and the poential of industrial fermentation gave rise to biotechnology.

The belief that the needs of a society could be met by fermenting industrial waste was an important ingredient of the "chemurgic movement," which began in the 1930s. Expectations were raised during the 1960s by a process that grew single-cell protein from petroleum oil. This new product was seen to be a solution to growing food shortages and world hunger. In the 1970s, biotechnology offered a possible solution to the escalation in oil prices and the increasing energy demands by synthesizing a new product called gasohol. However, both the single-cell protein and gasohol projects failed to progress due to varying issues including public resistance, a changing economic scene, and shifts in political power.

Yet the formation of a new field, genetic engineering, would soon bring biotechnology to the forefront of science in society, and the intimate relationship between the scientific community, the public, and the government would ensue. These debates gained exposure in 1975 at the Asilomar Conference, where Joshua Lederberg was the most outspoken supporter for this emerging field in biotechnology. By as early as 1978, with the synthesis of human insulin, Lederberg's claims would prove valid, and the biotechnology industry grew rapidly. Each new scientific advance became a media event designed to capture public support, and by the 1980s, biotechnology grew into a promising real industry. In 1988, only five proteins from genetically engineered cells had been approved as drugs by the FDA, but this number would skyrocket to over 125 by the end of the 1990s.

The field of genetic engineering remains a heated topic of discussion in today's society with the advent of gene therapy, stem cell research, cloning, and genetically-modified food. While it seems only natural nowadays to link pharmaceutical drugs as solutions to health and societal problems, this relationship of biotechnology serving social needs began centuries ago.

Origins of Biotechnology

Biotechnology arose from the field of zymotechnology, which began as a search for a better understanding of industrial fermentation, particularly beer. Beer was an important industrial, and not just social, commodity. In late 19th century Germany, brewing contributed as much to the gross national product as steel, and taxes on alcohol proved to be significant sources of revenue to the government. In the 1860s, institutes and remunerative consultancies were dedicated to the technology of brewing. The most famous was the private Carlsberg Institute, founded in 1875, which employed Emil Christian Hansen, who pioneered the pure yeast process for the reliable production of consistent beer. Less well known were private consultancies that advised the brewing industry. One of these, the Zymotechnic Institute, was established in Chicago by the German-born chemist John Ewald Siebel.

The heyday and expansion of zymotechnology came in World War I in response to industrial needs to support the war. Max Delbruck grew yeast on an immense scale during the war to meet 60 percent of Germany's animal feed needs. Compounds of another fermentation product, lactic acid, made up for a lack of hydraulic fluid, glycerol. On the Allied side the Russian chemist Chaim Weizmann used starch to eliminate Britain's shortage of acetone, a key raw material in explosives, by fermenting maize to acetone. The industrial potential of fermentation was outgrowing its traditional home in brewing, and "zymotechnology" soon gave way to "biotechnology."

With food shortages spreading and resources fading, some dreamed of a new industrial solution. Karl Ereky coined the word "biotechnology" in Hungary during 1917 to describe a technology based on converting raw materials into a more useful product. He built a slaughterhouse for a thousand pigs and also a fattening farm with space for 50,000 pigs, raising over 100,000 pigs a year. The enterprise was enormous, becoming one of the largest and most profitable meat and fat operations in the world. In a book entitled *Biotechnologie*, Ereky further developed a theme that would be reiterated through the 20th century: biotechnology could provide solutions to societal crises, such as food and energy shortages. For Ereky, the term "biotechnologie" indicated the process by which raw materials could be biologically upgraded into socially useful products.

This catchword spread quickly after the First World War, as "biotechnology" entered German dictionaries and was taken up abroad by business-hungry private consultancies as far away as the United States. In Chicago, for example, the coming of prohibition at the end of World War I encouraged biological industries to create opportunities for new fermentation products, in particular a market for nonalcoholic drinks. Emil Siebel, the son of the founder of the Zymotechnic Institute, broke away from his father's company to establish his own called the "Bureau of Biotechnology," which specifically offered expertise in fermented nonalcoholic drinks.

The belief that the needs of an industrial society could be met by fermenting agricultural waste was an important ingredient of the "chemurgic movement." Fermentation-based processes generated products of ever-growing utility. In the 1940s, penicillin was the most dramatic. While it was discovered in England, it was produced industrially in the U.S. using a deep fermentation process originally developed in Peoria, Illinois. The enormous profits and the public expectations penicillin engendered caused a radical shift in the standing of the pharmaceutical industry. Doctors used the phrase "miracle drug," and the historian of its wartime use, David Adams, has suggested that to the public penicillin represented the perfect health that went together with the car and the dream house of wartime American advertising. In the 1950s, steroids were synthesized using fermentation technology. In particular, cortisone promised the same revolutionary ability to change medicine as penicillin had.

Single-Cell Protein and Gasohol Projects

Even greater expectations of biotechnology were raised during the 1960s by a process that grew single-cell protein. When the so-called protein gap threatened world hunger, producing food locally by growing it from waste seemed to offer a solution. It was the possibilities of growing microorganisms on oil that captured the imagination of scientists, policy makers, and commerce. Major companies such as British Petroleum (BP) staked their futures on it. In 1962, BP built a pilot plant at Cap de Lavera in Southern France to publicize its product, Toprina. As there was no well-accepted term to describe the new foods, in 1966 the term "single-cell protein" (SCP) was coined at MIT to provide an acceptable and exciting new title, avoiding the unpleasant connotations of microbial or bacterial. By the late 1970s, however, the cultural climate had completely changed, as the growth in SCP interest had taken place against a shifting economic and cultural scene (136). First, the price of oil rose catastrophically in 1974, so that its cost per barrel was five times greater than it had been two years earlier. Second, despite continuing hunger around the world, anticipated demand also began to shift from humans to animals. The program had begun with the vision of growing food for Third World people, yet the product was instead launched as an animal food for the developed world. The rapidly rising demand for animal feed made that market appear economically more attractive. The ultimate downfall of the SCP project, however, came from public resistance. This was particularly vocal in Japan, where production came closest to fruition. For all their enthusiasm for innovation and traditional interest in microbiologically produced foods, the Japanese were the first to ban the production of single-cell proteins. The Japanese ultimately were unable to separate the idea of their new "natural" foods from the far from natural connotation of oil. These arguments were made against a background of suspicion of heavy industry in which anxiety over minute traces of petroleum was expressed. Thus, public resistance to an unnatural product led to the end of the SCP project as an attempt to solve world hunger.

In the late 1970s, biotechnology offered another possible solution to a societal crisis. The escalation in the price of oil in 1974 increased the cost of the Western world's energy tenfold. In response, the U.S. government promoted the production of gasohol, gasoline with 10 percent alcohol added, as an answer to the energy crisis. In 1979, when the Soviet Union sent troops to Afghanistan, the Carter administration cut off its supplies to agricultural produce in retaliation, creating a surplus of agriculture in the U.S. As a result, fermenting the agricultural surpluses to synthesize fuel seemed to be an economical solution to the shortage of oil threatened by the Iran-Iraq war. Before the new direction could be taken, however, the political wind changed again: the Reagan administration came to power in January 1981 and, with the declining oil prices of the 1980s, ended support for the gasohol industry before it was born.

Biotechnology seemed to be the solution for major social problems, including world hunger and energy crises. In the 1960s, radical measures would be needed to meet world starvation, and biotechnology seemed to provide an answer. However, the solutions proved to be too expensive and socially unacceptable, and solving world hunger through SCP food was dismissed. In the 1970s, the food crisis was succeeded by the energy crisis, and here too, biotechnology seemed to provide an answer. But once again, costs proved prohibitive as oil prices slumped in the 1980s. Thus, in practice, the implications of biotechnology were not fully realized in these situations. But this would soon change with the rise of genetic engineering.

Genetic Engineering

The origins of biotechnology culminated with the birth of genetic engineering. There were two key events that have come to be seen as scientific breakthroughs beginning the era that would unite genetics with biotechnology. One was the 1953 discovery of the structure of DNA, by Watson and Crick, and the other was the 1973 discovery by Cohen and Boyer of a recombinant DNA technique by which a section of DNA was cut from the plasmid of an E. coli bacterium and transferred into the DNA of another. This approach could, in principle, enable bacteria to adopt the genes and produce proteins of other organisms, including humans. Popularly referred to as "genetic engineering," it came to be defined as the basis of new biotechnology.

Genetic engineering proved to be a topic that thrust biotechnology into the public scene, and the interaction between scientists, politicians, and the public defined the work that was accomplished in this area. Technical developments during this time were revolutionary and at times frightening. In December 1967, the first heart transplant by Christian Barnard reminded the public that the physical identity of a person was becoming increasingly problematic. While poetic imagination had always seen the heart at the center of the soul, now there was the prospect of individuals being defined by other people's hearts. During the same month, Arthur Kornberg announced that he had managed to biochemically replicate a viral gene. "Life had been synthesized," said the head of the National Institutes of Health. Genetic engineering was now on the scientific agenda, as it was becoming possible to identify genetic characteristics with diseases such as beta thalassemia and sickle-cell anemia.

Responses to scientific achievements were colored by cultural skepticism. Scientists and their expertise were looked upon with suspicion. In 1968, an immensely popular work, *The Biological Time Bomb*, was written by the British journalist Gordon Rattray Taylor. The author's preface saw Kornberg's discovery of replicating a viral gene as a route to lethal doomsday bugs. The publisher's blurb for the book warned that within ten years, "You may marry a semi-artificial man or woman...choose your children's sex...tune out pain...change your memories...and live to be 150 if the scientific revolution doesn't destroy us first." The book ended with a chapter called "The Future – If Any." While it is rare for current science to be represented in the movies, in this period of "Star Trek", science fiction and science fact seemed to be converging. "Cloning" became a popular word in the media. Woody Allen satirized the cloning of a person from a nose in his 1973 movie *Sleeper*, and cloning Hitler from surviving cells was the theme of the 1976 novel by Ira Levin, *The Boys from Brazil*.

In response to these public concerns, scientists, industry, and governments increasingly linked the power of recombinant DNA to the immensely practical functions that biotechnology promised. One of the key scientific figures that attempted to highlight the promising aspects of genetic engineering was Joshua Lederberg, a Stanford professor and Nobel laureate. While in the 1960s "genetic engineering" described eugenics and work involving the manipulation of the human genome, Lederberg stressed research that would involve microbes instead. Lederberg emphasized the importance of focusing on curing living people. Lederberg's 1963 paper, "Biological Future of Man" suggested that, while molecular biology might one day make it possible to change the human genotype, "what we have overlooked is euphenics, the engineering of human development." Lederberg constructed the word "euphenics" to emphasize changing the phenotype after conception rather than the genotype which would affect future generations.

With the discovery of recombinant DNA by Cohen and Boyer in 1973, the idea that genetic engineering would have major human and societal consequences was born. In July 1974, a group of eminent molecular biologists headed by Paul Berg wrote to *Science* suggesting that the consequences of this work were so potentially destructive that there should be a pause until its implications had been thought through. This suggestion was explored at a meeting in February 1975 at California's Monterey Peninsula, forever immortalized by the location, Asilomar. Its historic outcome was an unprecedented call for a halt in research until it could be regulated in such a way that the public need not be anxious, and it led to a 16-month moratorium until NIH guidelines were established.

Lederberg was the leading exception in emphasizing, as he had for years, the potential benefits. At Asilomar, in an atmosphere favoring control and regulation, he circulated a paper countering the pessimism and fears of misuses with the benefits conferred by successful use. He described "an early chance for a technology of untold importance for diagnostic and therapeutic medicine: the ready production of an unlimited variety of human proteins. Analogous applications may be foreseen in fermentation process for cheaply manufacturing essential nutrients, and in the improvement of microbes for the production of antibiotics and of special industrial chemicals." In June 1976, the 16-month moratorium on research expired with the Director's Advisory Committee (DAC) publication of the NIH guidelines of good practice. They defined the risks of certain kinds of experiments and the appropriate physical conditions for their pursuit, as well as a list of things too dangerous to perform at all. Moreover, modified organisms were not to be tested outside the confines of a laboratory or allowed into the environment.

Atypical as Lederberg was at Asilomar, his optimistic vision of genetic engineering would soon lead to the development of the biotechnology industry. Over the next two years, as public concern over the dangers of recombinant DNA research grew, so too did interest in its technical and practical applications. Curing genetic diseases remained in the realms of science fiction, but it appeared that producing human proteins would be good business. Insulin, one of the smaller and best understood proteins, had been used in treating diabetes for half a century. It had been extracted from animals in a chemically slightly different form from the human product. Yet, if one could produce human insulin, one could meet an existing enormous demand with an apparently superior product whose approval would be relatively easy. In the period 1975 to 1977, insulin represented the aspirations for new products that could be made with the new biotechnology. Microbial production of human insulin was finally announced in September 1978 and was commercially produced by a startup company, Genentech.

The radical shift in the connotation of "genetic engineering" from an emphasis on the inherited characteristics of people to the commercial production of proteins and therapeutic drugs was nurtured by Joshua Lederberg. His broad concerns since the 1960s had been stimulated by enthusiasm for science and its potential medical benefits. Countering calls for strict regulation, he expressed a vision of potential utility. Against a belief that new techniques would entail unmentionable and uncontrollable consequences for humanity and the environment, a growing consensus on the economic value of recombinant DNA emerged.

Biotechnology and Industry

With ancestral roots in industrial microbiology that date back centuries, the new biotechnology industry grew rapidly beginning in the mid-1970s. Each new scientific advance became a media event designed to capture investment confidence and public support. Although market expectations and social benefits of new products were frequently overstated, many people were prepared to see genetic engineering as the next great advance in technological progress. By the 1980s, biotechnology characterized a nascent real industry, providing titles for emerging trade organizations such as the Industrial Biotechnology Association.

The main foci of attention after insulin were the potential profit makers in the pharmaceutical industry: human growth hormone and what promised to be a miraculous cure for viral diseases, interferon. Cancer was a central target in the 1970s because increasingly the disease was linked to viruses. By 1980, a new company, Biogen, had produced interferon through recombinant DNA. The emergence of interferon and the possibility of curing cancer raised money in the community for research and increased the enthusiasm of an otherwise uncertain and tentative society. Moreover, to the 1970s plight of cancer was added AIDS in the 1980s, offering an enormous potential market for a successful therapy, and more immediately, a market for diagnostic tests based on monoclonal antibodies. By 1988, only five proteins from genetically engineered cells had been approved as drugs by the FDA: insulin, human growth hormone, hepatitis B vaccine, alpha-interferon, and tissue plasminogen activator (TPa), for lysis of blood clots. By the end of the 1990s, however, 125 more genetically engineered drugs would be approved.

Genetic engineering also reached the agricultural front as well. There was tremendous progress since the market introduction of the genetically engineered Flavr Savr tomato in 1994. Ernst and Young reported that in 1998, 30% of the U.S. soybean crop was expected to be from genetically engineered seeds. In 1998, about 30% of the US cotton and corn crops were also expected to be products of genetic engineering.

Genetic engineering in biotechnology stimulated hopes for both therapeutic proteins, drugs and biological organisms themselves, such as seeds, pesticides, engineered yeasts, and modified human cells for treating genetic diseases. From the perspective of its commercial promoters, scientific breakthroughs, industrial commitment, and official support were finally coming together, and biotechnology became a normal part of business. No longer were the proponents for the economic and technological significance of biotechnology the iconoclasts. Their message had finally become accepted and incorporated into the policies of governments and industry.

Global Trends

According to Burrill and Company, an industry investment bank, over $350 billion has been invested in biotech so far, and global revenues have risen from $23 billion in 2000 to more than $50 billion in 2005. The greatest growth has been in Latin America but all regions of the world have shown strong growth trends.

There has been little innovation in the traditional pharmaceutical industry over the past decade and biopharmaceuticals are now achieving the fastest rates of growth against this background, particularly in breast cancer treatment. Biopharmaceuticals typically treat sub-sets of the total population with a disease whereas traditional drugs are developed to treat the population as a whole. However, one of the great difficulties with traditional drugs are the toxic side effects the incidence of which can be unpredictable in individual patients.

PLANT BIOTECHNOLOGY

Plant biotechnology offers several possibilities for increasing productivity, diversification and production, while developing a more sustainable agriculture (Izquierdo, 1999). This technology includes biopesticide production, plant tissue culture techniques, and the use of advanced molecular biology techniques for plant transformation, genomic analysis coupled with breeding and plant-disease diagnoses. It can be an effective instrument to mitigate the consequences of climatic change and can offer cropping alternatives in lands degraded by erosion and desertification or by careless agricultural use (FAO, 1999). Agricultural biotechnology has different levels of implementation and development according to its degree of complexity, infrastructure costs and investment.

Plant breeding is the purposeful manipulation of plant species in order to create desired genotypes and phenotypes for specific purposes. This manipulation involves either controlled pollination, genetic engineering, or both, followed by artificial selection of progeny. *Plant breeding* often, but not always, leads to plant domestication.

Plant breeding has been practiced for thousands of years, since near the beginning of human civilization. It is now practiced worldwide by government institutions and commercial enterprises. International development agencies believe that breeding new crops is important for ensuring food security and developing practices of sustainable agriculture through the development of crops suitable for their environment

The first level includes technologies of low complexity and cost like handmade and semi-industrial production of biological products, like bio-inoculants for leguminous crops, reproduction of pest antagonist and pest predators and production of bio-pesticides. The transfer of these simple technologies to the small producers has proven feasible and it can offer remarkable results quickly.

The second level comprises the application of cell and tissue culture techniques, cell culture for production of secondary metabolites, disease diagnostic systems based on polyclonal antibodies and other technologies with medium complexity. This stage can also include investments to implement semi-industrial production of biological products. The results of these technologies can be seen in the medium term (three to five years), depending of the complexity of the projects and the degree of organisation for extending the results to the field or laboratory. The first two stages do not require many resources but an efficient organisation for the transfer of technology to extend the results to the producers in a brief period of time.

The third level considers the use of molecular biology, genetic engineering, large-scale industrial production of biological products and complex plant cell and tissue culture techniques like liquid-bioreactors. This level is characterised by its complexity and high costs of research and development. The short term impact of this advanced biotechnology is limited because, as in any novel technology, it needs a starting phase corresponding to the innovation of its own research methods, analytical techniques and a semi-industrial base to supply the needed inputs and equipment. To create such conditions it is compulsory to build a framework of institutional support by public and private investment sectors.

Basic knowledge on the complexity of the molecular organisation of living organisms and their interrelationships is far from its complete elucidation and limits the number of characters that can be manipulated by genetic engineering. Identification techniques, isolation and characterisation of new genes to be transferred to different crops to produce genetically modified plants (GMP) are subjected to constant research.

Plant genetic engineering now permits manipulating monogenic characters while the productive characteristics of the crop species, of a quantitative nature result in the expression of multiple genes (polygenic). Often, one isolated gene does not have a great influence in the complete genome expression and this reinforces the importance of conventional plant breeding assisted by molecular markers (Robentson, 1989). The above issues are related to proprietary rights and institutional capacity building.

The right of farmers to freely reuse a genetically modified food crop variety to strengthen food security and the patenting and marketing of GMP´seeds, is under dispute. Biotechnology has begun to have an economic impact in the productivity of crops and in the decrease of production costs. In 1996 in the United States 15 novel GMP products existed in the market whose initial sales totalled about 380 million dollars. It is considered that this emergent market will increase 20% annually during the next decade (Persley, 1997).

In that year, GMP were planted for the first time in Latin America comprising commercial fields, among which are herbicide tolerant soybean and insect resistant corn and cotton (Woodson, 1997) introduced by transnational companies after a case-to-case approval by national authorities. In 1999, 39.9 million hectares were globally planted, 6.7 in Argentina alone (James, 1999), resulting in a 44% increase in the adoption of such a technology by the medium and large farms. The release of genetically modified organisms is a matter of growing concern and requires the evaluation of biological risks and their environmental and socio-economic impact together with appropriate technological packages (FAO, 1999).

Modern Plant Breeding

Modern plant breeding uses techniques of molecular biology to select, or in the case of genetic modification, to insert, desirable traits into plants.

Marker-Assisted Selection

Sometimes many different genes can influence a desirable trait in plant breeding. The use of tools such as molecular markers or DNA fingerprinting can map thousands of genes. This allows plant breeders to screen large populations of plants for those that possess the trait of interest. THe screening is based on the presence or absence of a certain gene as determined by laboratory proceedures, rather than on the visual identification of the expressed trait in the plant.

Genetic Modification

Genetic modification of plants is achieved by adding a specific gene or genes to a plant, or by knocking out a gene with RNAi, to produce a desirable phenotype. The plants resulting from adding a gene are often referred to as transgenic plants. If for genetic modification genes of the species or of a crossable plant are used under control of their native promoter, then they are called Cisgenic plants. Genetic modification can produce a plant with the desired trait or traits faster than classical breeding because the majority of the plant's genome is not altered.

To genetically modify a plant, a genetic construct must be designed so that the gene to be added or knocked-out will be expressed by the plant. To do this, a promoter to drive transcription and a termination sequence to stop transcription of the new gene, and the gene of genes of interest must be introduced to the plant. A marker for the selection of transformed plants is also included. In the laboratory, antibiotic resistance is a commonly used marker: plants that have been successfully transformed will grow on media containing antibiotics; plants that have not been transformed will die. In some instances markers for selection are removed by backcrossing with the parent plant prior to commercial release.

The construct can be inserted in the plant genome by genetic recombination using the bacteria *Agrobacterium tumefaciens* or *A. rhizogenes*, or by direct methods like the gene gun or microinjection. Using plant viruses to insert genetic constructs into plants is also a possibility, but the technique is limited by the host range of the virus. For example, Cauliflower mosaic virus (CaMV) only infects cauliflower and related species. Another limitation of viral vectors is that the virus is not usually passed on the progeny, so every plant has to be inoculated.

The majority of commercially released transgenic plants, are currently limited to plants that have introduced resistance to insect pests and herbicides. Insect resistance is achieved through incorporation of a gene from *Bacillus thuringiensis* (Bt) that encodes a protein that is toxic to some insects. For example, the cotton bollworm, a common cotton pest, feeds on Bt cotton it will ingest the toxin and die. Herbicides usually work by binding to certain plant enzymes and inhibiting their action. The enzymes that the herbicide inhibits are known as the herbicides *target site*. Herbicide resistance can be engineered into crops by expressing a version of *target site* protein that is not inhibited by the herbicide. This is the method used to produce glyphosate resistant crop plants (See Glyphosate)

Genetic modification of plants that can produce pharmaceuticals (and industrial chemicals), sometimes called *pharmacrops*, is a rather radical new area of plant breeding.

Agricultural Biotechnology in the Region

Agricultural biotechnology development in the LAC Region is characterised by a repetitive and academic model in where many of the project objectives do not respond to the real needs of crop production and food security in the Region. Frequently, research projects and laboratories repeat work that has already been conducted by advanced research groups of developed countries without any further innovation and with a low degree of scientific creativity.

On consulting the 30 scientific publications with the highest impact on biotechnology, we find that only 2% of the published scientific articles were generated in laboratories of the Region (data from Current Contents, 1997 processed by authors). The regional situation was surveyed in 1990 by the Technical Co-operation Network on Plant Biotechnology (REDBIO/FAO), where 63.4% of the plant biotechnology activities were focused on tissue culture with the support of the private sector, 10.1% corresponded to projects on plant disease diagnoses while genetic engineering and functional genomic analysis represented 8.6 and 2.7% respectively.

Optimistically, the survey allowed to calculate for the next decade an estimated growth of 22.6% in the genetic engineering activities and of 4.1% in genomic analysis (Izquierdo et al.1995). In 1997, another REDBIO survey revealed that tissue culture maintained its predominance with 51.7% due to its immediate application in agriculture. An interesting aspect found was the increase of project research on genomic analysis, although not at the same scientific level in all laboratories. This responds to the need to complement plant breeding projects assisted by molecular markers and to the difficulties found with somaclonal variation (4.9% from an estimate of 12.6%). The expected expansion in genetic engineering did not take place; in fact genetic engineering fell in its relative position from 8.6 % in 1990 to 6.4% in 1997. This tendency is the result of the complexities of these techniques, the high cost and the already mentioned limitations of genetic engineering at present. Many small research teams in universities or agricultural institutions, poorly connected and/or integrated, have a high dispersion of facilities and qualified labor force. According to the database of REDBIO/FAO in 1999, 539 laboratories are affiliated in 23 countries of the Region. 83% of the member laboratories have less than 10 researchers and technical personnel and only 72% of them have three postgraduates or more.

These small research groups lack adequate technology and trained human resources to start relevant projects attractive for public and private investment. This situation reduces opportunities for the development of projects on plant biotechnology with a significative impact on crop production or food security. In spite of this, 48% of the laboratories are directly related to private investment and produce healthy plants through tissue culture and micropropagation, offering a supply of about 75 million plants annually. Research and development in other areas of biotechnology such as bioinsecticide production, inoculants, entomophagues and entomopathogens was increased. These activities, previously defined as parts of the first development stage, are able to be applied in the agriculture sector and to report short-term benefits.

The number of investigations related to the study of the secondary pathway compounds of medicinal plants and their therapeutic use for further research on molecular farming also increased. These projects have promising perspectives, but they require an appropriate analytic basis and an infrastructure that should grow as on-going projects move to upper stages of development with a reliable medical validation of these products. This process requires the application of strict norms of quality control, registration and medical approval of new products for pharmaceutical use. The range of species used in research and development projects also changed during the last decade. In 1990, 29% of the activities were directed to fruit-bearing plants, 28% to roots and tubers, and only 1.1% to ornamental plants.

In 1997 the fruit-bearing plants represented most of the activities (27%), activities on roots and tubers decreased to 14% while in ornamental plants rose to 10%. Economic reasons led to those changes. Propagation and tissue culture of fruit-bearing and ornamental plants are more profitable due to the high commercial value of the propagated plants. On the other hand, in species more important for food security, like the edible roots and tuber plants, the price of the propagated plant should go down because its actual cost is very high, limiting its access by small producers. This is a factor that affects the impact of plant biotechnology on food production seriously. Genetic engineering provides powerful tools to enhance the modification of plants to the potential benefit of society. However, as with any new scientific advancement, careful consideration of the effects of employing these tools is necessary to ensure that the result will be a net benefit to society.

Recent controversies about genetically engineered crops have highlighted the need for experimental evidence and sound scientific judgment to assess the risks versus benefits. This debate was once relegated mainly to the plant scientists and activists and focused only on the food safety aspects. It has now flowed into the realm of the biomedical sciences with issues such as predicting allergenicity, assessing nutritional benefit, evaluating nutritional quality, meeting the nutritional needs of developing nations, and expanding the sustainable food supply to meet future demands. Pant biologists and animal biologists often conduct their inquiries in parallel universes; interactions between the two fields of science are serendipitous and unsystematic. The Life Sciences Research Office held a forum during Experimental Biology 2001 to present current topics in food biotechnology to the experimental biology community with the hope of bridging the two universes. The conference had the further goal of identifying those areas in need of future research. This paper summarizes the benefits and risks of this new technology, describes consumers' knowledge and attitudes, explains the regulatory process new products of biotechnology go through prior to commercialization, and identifies challenges facing the industry, consumers and regulators.

Agricultural biotechnology may be defined as the use of living plant organisms, or parts thereof, to produce food and feed products such as insect-resistant corn, to develop processes like the manufacturing of biologics by tobacco, and to provide services such as, bioremediation of heavy metal contamination using genetically engineered poplars. Although biotechnology appears to be a new technology, the underlying concept is not new. Farmers have been using genetic manipulation to improve crops for thousands of years. For example, some 8000 years ago the Native Americans created corn by domestication of a wild plant called teosinte. Teosinte has a short, thin ear with very small kernels. The Native Americans used selective breeding, a crude form of genetic manipulation, in a remarkable way to produce a more productive variety. The end product looks very similar to the varieties of corn we produce today.

During the last century plant breeders expanded the tools of genetic manipulation beyond conventional cross breeding to use a variety of other breeding techniques, including embryo rescue, chemical mutagenesis, radiation mutagenesis, and somaclonal variation. These techniques do not allow control at the genome level; rather they allow multiple genes to transfer and require a rigorous selection process to ensure that the desired characteristic is stable. Plants created by these conventional phenotypic selection techniques do not undergo formal food or environmental safety evaluation prior to introduction into the environment and marketplace. On the other hand, over the past 20 years, the development of genetic engineering techniques now allows the development of crops containing specific single gene transfers. These are controllable, testable, and predictable changes, grounded on scientific principle. Genetically engineered crops undergo extensive testing of composition, safety, agronomic traits, and environmental effects prior to introduction into the marketplace. These assessments are described further in the section of this paper, entitled Regulation of Crop Biotechnology.

Benefits

Modifications of crop plants can be organized into two main categories: those that benefit the producer and those that benefit the consumer. Modifications that protect the crop from either biotic or abiotic stress or increase total crop yield benefit the producer and are called input traits. The majority of modified crops in commercial use fit in this group. Scientists have just begun to tap the large potential of biotechnology to produce varieties of plants that confer a wide spectrum of advantages to consumers. These varieties are modified with output traits. One of the most publicized uses of biotechnology in agriculture is the modification of corn to express proteins produced by the common soil bacterium, *Bacillus thuringiensis* (Bt). Organic farmers have been using Bt as an insecticidal spray for over 40 y. Bt organisms have been modified to express a class of insecticidal proteins called Cry. These proteins are effective against certain insect pests but are harmless to humans, mammals and birds. Bt corn was introduced as a commercial crop in 1996 and has been described as "the most important technological advancement in insect pest management since the development of synthetic insecticides" because of its inherent resistance to infestation by one of the most serious corn pests, the European corn borer (*Ostrinia nubilalis*). In addition to decreasing yields, infestation of corn by the European corn borer facilitates spoilage by the mold, fusarium, which forms a mycotoxin, fumonosin, in corn. Fumonosin is a toxic substance that, among other things, has produced liver damage in all animals studied.

Although currently inconclusive, some evidence suggests that it may also play a role in human esophageal cancer. Studies challenging maize hybrids with the European corn borer found decreased fumonisin concentrations for transgenic maize varieties expressing specific Cry proteins, 2.1 ug/g compared to 16.5 ug/g for nontransgenic maize hybrids. Beyond the health benefit to livestock and consumers, infected corn becomes an economic problem. Corn that exceeds levels allowable for the intended use must either be discarded or used for another purpose, causing a loss of profit.

In fact, overall losses to farmers as a result of the European corn borer-infested crops (including corn, cotton, sorghum, and other vegetables) total 1 billion dollars a year. On the whole, Bt corn and Bt cotton have improved farm efficiency. A report by the U.S. Department of Agriculture concludes by saying that even though the benefit and performance of genetically modified crops varies depending on many factors including region and pest infestation levels, the adoption of crops such as Bt cotton in the Southwest and herbicide-tolerant soybeans led to significant increases in net yield, and a significant decrease in the application of insecticide. An expansion of the same report by the Economic Research Service investigating same-year differences between average pesticide use of adopting and nonadopting farmers showed that those using genetically modified corn, soybeans and cotton combined used 7.6 million fewer acre-treatments (acres being treated multiplied by the number of treatments) of pesticides compared to nonadopters in 1997.

In 1998 the difference increased to almost 17 million fewer acre-treatments by adopters. On the other hand, others have concluded that despite a sharp increase in planting Bt corn, the percentage of corn treated with insecticides has remained constant throughout the period. Moreover, some research suggested that Bt corn pollen was harmful to the Monarch butterfly. Collaborative research by scientists from universities and research institutions in the US and Canada have concluded that potential risks of Bt corn to Monarchs is low.

This is due to the density and time of pollen shed in relation to the period of larvae activity. Also, the proportion of milkweed plants growing near cornfields and the proportion of fields that are planted with Bt corn are other mitigating factors cited for the low risk to Monarch butterflies. Other traits have been added to a variety of crops to defend from biological insults. Tomato, potato, squash and papaya are among a variety of crops that have been modified to resist infection by viruses or insect pests. In addition to biotic stressors, plant productivity is influenced by abiotic factors such as herbicides, soil composition, water supply, and temperature.

Therefore, conferring plants with genes that will help them withstand a wider range of environmental conditions could increase productivity. Plants are also being engineered to withstand drought, heat, cold temperatures and poor soil conditions such as salinity and aluminum contamination. Increased total yield of harvest also can be achieved by enhancing efficiencies in the metabolic and photosynthetic pathways. Examples of pathways that could be improved to increase crop yield include nitrogen assimilation, starch biosynthesis, and modification of photosynthesis. After harvest, time to market is an important economic factor due to the perishability of produce. Changing the rate of ripening would seem to be a benefit to both the farmer, by decreasing post harvest losses, and the consumer by increasing shelf life.

To prevent delivering spoiled fruit, mature tomatoes are harvested while still green and ripened during delivery by exposure to ethylene, a ripening hormone in tomatoes. In 1994 the Food and Drug Administration approved a brand of tomato that had a genetic solution to this processing problem. The producers of the tomato used antisense technology to silence a gene that produces polygalacturonase, a pectin-degrading enzyme found in ripe tomatoes, thus slowing the ripening process. Consumers stand to gain more than just produce with longer shelf lives. While still in its infancy, the technology is being used to produce plants that will have a whole range of output traits including increased nutritional value, medicinal properties, industrial utility, and novel taste and esthetic appeal. Many of our common food crops could be improved to better meet the nutritional requirements of humans or animals. Protein, starch and oil composition and content as well as micronutrient content can all be improved to make foods and feeds more nutritious.

For example, a new strain of potatoes containing 30–60% more starch has been developed by inserting a bacterial gene for an enzyme in the starch biosynthetic pathway. These high-starch potatoes have less moisture and therefore absorb less fat during frying. Enzyme biotechnology also is being used to develop specialty oils containing more favorable fatty acid profiles such as high oleic acid peanut oil. We have known for a long time that vitamins and minerals elicit biologic responses and have positive effects on health. Carotenoids are another class of nutrients that may be associated with decreased risk for certain cancers and macular degeneration.

Among other plants they can be found in papaya (ß-carotene), tomatoes (lycopene), kale and spinach (lutein). Beta-carotene has already been expressed in a genetically engineered rice cultivar, named Golden Rice, by addition of genes for three enzymes in the phytoene synthase pathway (two genes from a daffodil and one from the bacteria Erwinia uredovora). This strain was also crossed with a high iron strain of rice to produce a strain with both qualities. Golden Rice has been the subject of much attention because it represents the potential of future biotechnology crops to benefit people in developing countries. This variety of rice could decrease malnutrition and blindness associated with vitamin A deficiency. However, there have been questions raised about the effectiveness of this rice because of the many biological, cultural and dietary barriers that must be overcome. These questions will need to be answered as the product is further developed prior to introduction. Other phytonutrients with purported health benefits include glucosinolates, phytoestrogens, and phytosterols.

Found in a wide variety of food sources, these compounds could selectively be overexpressed to therapeutic levels. In a much different way, biotechnology is poised to completely change another aspect of preventative health care. Methods are being developed to produce vaccines in plants by introducing genes that express a protein antigen in crops such as corn, potatoes and bananas. When eaten these antigens elicit an immune response and have been shown to provide protection against a subsequent challenge from pathogens. The feasibility of this approach was demonstrated when mice fed transgenic Hepatitis B surface antigen expressed in potato tubers showed a primary immune response by producing antibody specific to the antigen.

Companies are positioning themselves to become suppliers of a wide range of biotechnology products, including bioactive therapeutic proteins, blood proteins, animal health products, and industrial enzymes. There are many other possible industrial applications for genetically modified organisms. For example, researchers at the University of Georgia engineered yellow poplar trees to have the ability to extract toxic ionic mercury from soil and convert the toxin to a relatively inert form. The gene was acquired from mercury-resistant bacteria that are soil-borne and thrive at sites polluted with heavy metals. In one study the engineered plants were capable of ten times the rate of mercury removal as compared to nonengineered plants.

This is just one example of how phytoremediation, the use of plants to clean or contain contaminated areas, combined with biotechnology is a promisingly efficient, economical, and environmentally friendly technique that could restore soil health and revegetate contaminated waste sites. Clearly, consumer preference is playing a role in the industry's choice of product development. Just as consumers have appreciated seedless oranges and watermelon, the industry is developing other fruiting crops that do not require fertilization to produce seedless fruit. Novel produce such as seedless tomatoes, squash, eggplant, peppers, strawberries, melons and cherries, to list a few of the possibilities, would be attractive to many consumers. Additionally, these fruits would have improved taste due to increased total soluble sugar compared to seeded fruit and may be of economic value to the processing industry as well.

Also in development is a sweet protein found naturally in the fruit of the African vine, Pentadiplandra brazzeana. This heat stable protein is 500 times as sweet as sucrose at higher concentration and as much as 2000 times as sweet in a two percent (by weight) solution. Lacking bitterness, it has a lot of potential as an alternative low energy sweetener.

It is clear that plant biotechnology has the potential to have a huge affect on tomorrow's society. Already over 50 biotechnology crop products have passed the regulatory review process and have been commercialized, ranging from corn and potatoes to tomatoes and squash. It is interesting to note that crops modified by biotechnology are the most rapidly adopted technology in the history of agriculture. In 1996 only 4.3 million acres of biotech crops had been planted; by 2000 that number increased to 109.2 million acres.

Plant Biotechnology Regulation

The regulatory process for the use of any technology should take into account both the risks and benefits of implementing that technology as well as the consequences of nonimplementation and the viability of alternative technologies. Biotechnology poses unique challenges to the regulatory process because it is a new tool with the potential to affect a wide spectrum of changes in an equally wide range of products.

Therefore, each use of the technology must be reviewed on a case-by-case basis. Blanket acceptance or rejection of this technology will take years of practical experience.

Four federal entities play different roles in the current process, the National Institutes of Health (NIH), the Animal Plant Health Inspection Service (APHIS) of the USDA, the Food and Drug Administration (FDA), and the Environmental Protection Agency (EPA). The combination of the regulatory protocols provides coverage over all aspects of plant biotechnology from the laboratory to the farm to the marketplace.

The first step in making a product is at the laboratory research level. The NIH has published a set of guidelines to specify practices for working with recombinant DNA and organisms and viruses containing recombinant DNA to promote a high level of biosafety. The NIH Guidelines for Research Involving Recombinant DNA Molecules cover all aspects of recombinant DNA research and require compliance from any institution that receives funds from the NIH. These guidelines are widely accepted and followed, even by institutions that do not receive NIH funding. Under the guidelines, institutions that conduct this type of research must set up an Institution Biosafety Committee (IBC) that is composed of experts in various fields relating to the subject matter and are encouraged to open their meetings to the public. The IBC is responsible for all the guidelines set forth by the NIH and must approve a project either before or during initiation, depending of the nature of the project and is keenly aware of pathogenicity and containment issues.

According to the Federal Plant Protection Act (FPPA), any transgenic crop containing DNA of a known plant pest is viewed as a potential plant pest and is therefore under the regulatory control of the APHIS Plant Protection and Quarantine within the USDA. Crop developers must file a notification for permission to import, grow, or move through state lines any genetically engineered plants. Developers must meet several criteria for APHIS to accept the notification. The species of the crop must already be considered safe for growth. The introduced genetic material must be stable. The function of the new genetic material must be known, and its expression in the regulated article cannot result in plant disease, produce products intended for pharmaceutical use, produce an infectious entity, or be toxic to a nontarget organism. In addition, the added sequence cannot pose a risk of creating any new viruses or be derived from any animal or human pathogen. If approved, the organism can only be grown with proper containment for data collection purposes and must be destroyed afterwards. After the developer has collected sufficient data, APHIS accepts petitions to deregulate the crop. Public input is invited at this stage. If found genetically stable and there is no finding of environmental concerns, the USDA grants a permit and no longer regulates the crop.

Mandated by the Federal Food, Drug, and Cosmetic Act (FFDCA) and the Federal

Insecticide, Fungicide, and Rodenticide Act (FIFRA), the EPA is the regulatory agency ensuring the health and environmental safety of all pesticides. Therefore, the EPA becomes involved in the review of biotechnology-derived crops when they are altered to produce a pesticide (plant-incorporated protectant or PIP). To register PIPs for commercial use with the EPA, the developer must submit an extensive amount of data concerning the toxicity, biodegradability, environmental effect, and effects on nontarget organisms. The EPA Office of Pesticide Programs grants an Experimental Use Permit allowing for planting in order to conduct field trials for the purpose of collecting the Agency's required data. Public comment is invited throughout each stage of this process via publication in the Federal Register. Under FIFRA, the EPA must review all data concerning the ecological, human, and animal health effects from the use of the PIP to certify that there will be no unreasonable adverse human or ecological effects. In addition, under the Food Quality Protection Act (FQPA), EPA must set safe tolerances for PIPs in foods for human consumption. It is then the role of the FDA to enforce the tolerance level in foods postproduction. The EPA may grant an exemption from a tolerance if there are extensive data on the safety of the introduced trait.

The FDA's mission, among other things, is to protect public health by ensuring the safety of foods. The FDA currently does not have a mandatory premarket approval system in place for food products derived from biotechnology because it does not consider *the method of production* to raise significant safety issues. Rather, the FDA treats a food developed via biotechnology like other new foods and bases the safety assessment on the premise that a new food should be "as safe as" the food from which it was developed. The principles of safety assessment were published in the FDA's 1992 policy statement, Foods Derived From New Plant Varieties.

The FDA concluded that it is sensible for developers of new varieties of foods to consult with the FDA so that "relevant scientific, safety, and regulatory issues are resolved prior to the introduction of such products into the marketplace". To facilitate this consultation process, the Biotechnology Evaluation Team (BET) was created by the FDA's Center for Food Safety and Applied Nutrition (CFSAN), Office of Surveillance and Compliance (OSC), and Center for Veterinary Medicine (CVM). The BET is composed of a consumer safety officer, molecular biologist, chemist, environmental scientist, toxicologist, nutritionist, and may include supplemental expertise depending on the product in question. The BET requests information, such as intended uses of the food, source of genetic material, intended technical affect of the modification, the expected effect of the composition of the food, the identity, function, and concentration of all newly expressed products, any known or suspected allergens or toxins, a comparison of the composition of the new food to that of the parental variety, and any other relevant information concerning the safety and nutritional assessment of the bioengineered food. Developing this information can require several years.

Although the FDA's premarket review process has been voluntary, every company that has commercialized a new product developed via biotechnology has submitted its data to FDA for review, a process that can take several years. The FDA recently proposed making this premarket notification process mandatory and is currently receiving comments on the matter.

Throughout the food regulatory system the goal is minimizing risks by ensuring that a new technology or product is "as safe as" a similar product before it. Consequently the assessments are always comparative. A hazard is a hypothetical event that might do harm, and a risk is a hazard that actually occurs at some known frequency or probability. Food safety is achieved by risk identification, comprehensive hazard analysis, assessment, concluding that benefits outweigh costs, development of a risk management plan, and effective training and risk communication. Using all of these elements in a food safety evaluation maintains a precautionary approach.

Foods derived from biotechnology are more heavily regulated than any other new foods. Only new food additives and food ingredients are subjected to premarket approval— new nongenetically modified whole foods are not—and there is no required premarket screening of whole foods for toxicity or allergenicity.

Aspects of our food supply that may pose health risks include food borne illness, untested natural foods, dietary supplements, and dietary inadequacies such as over-nutrition, under-nutrition, and nutrition inadequacy. There are natural toxicants, allergies to various foods, chance additives, pesticide and herbicide residues, food ingredients and additives. Biotechnology carries a relatively low risk to human health and, in fact may become part of the solution to some of the problems mentioned above.

Consumers need to realize that there is extensive safety assessment and regulatory evaluation done on these products before commercialization. However, a fundamental difference in opinion that is driving controversy needs to be resolved: Are transgenics inherently more or less safe? For this we need additional scientific evidence and critical thinking that would lead to science-driven policy decisions. We should learn to view safety as a comparative finding, where the safety of a product from biotechnology is determined in relation to the safety of the parent product. Scientists and regulators need to build trust and credibility with consumers without using the "trust me" approach and should strive to be better educators because of the many challenges ahead in terms of acceptance for this technology.

Challenges for Industry, Regulators and Consumers

New technologies take time for acceptance. For crop biotechnology, key factors hindering acceptance are the public's limited understanding of modern agricultural practices and the science involved in biotechnology. Food, an emotional and personal topic, combined with misunderstanding of biotechnology, sensationalized media coverage, and complex ethical and social matters have combined to generate fear in some consumers. Well-organized protest groups will likely continue to center the public debate on issues of uncertainty and mistrust.

Yet biotechnology is becoming more and more a part of our lives. Genome sequencing, and analysis using new tools such as microarrays, allowing for very high throughput analysis of genes and gene expression patterns, are accelerating the discovery of genes with possible medical utility. Due to discovered homology in the human genome with that of other species, the use of nonhuman genes and proteins in human therapeutics will become more common. Revenues are expected to increase for biotechnology-modified crops in general, as a result of the production of crops containing "output traits."

One of the biggest barriers facing the acceptance of this technology are European import restrictions. Antibiotechnology sentiment is much stronger in Europe than in the U.S. European public support for applications of biotechnology that would protect crops from insects dropped from 58% in 1996 to 42% in 1999. Support for biotechnology-improved foods also dropped from 44% in 1996 to 31% in 1999. There are many social and cultural reasons for this increase in opposition in Europe, including more sensational and negative media coverage, general support for small farms and open spaces, and a general opposition to foreign producers and their products. There are also political and economical reasons for greater opposition in Europe. Past food crises (e.g., BSE) have undermined confidence in government and exacerbated a general paranoia concerning the food supply. There has been limited educational outreach by the food industry, public leaders, academia or consumers. Antiglobalization sentiments and nontariff trade barriers have further complicated the issue.

Taken as a whole, the European situation is very complicated and unpredictable. However, many key leaders seem to want resolution, and as more products produced using this technology arrive on the market from around the world, there will be fewer alternatives. Consumer education could also be effective if led by European leaders and scientists. So far, the European Union has set out a plan for genetically modified food labels. It states that labels are required if any ingredient in a product contains more than 1% of GM material. Afraid that consumers view GM labels as a warning, European manufacturers are avoiding GM ingredients, thus eliminating choice at the market.

European consumers have expressed interest in identity preservation of GM foods. Undeniably, identity preservation adds cost and complexity to the system. With so many different steps from seed to market, it would take countless measures to ensure proper labeling. This is further complicated by the fact that these genes may not be absolutely confined. For example, it is estimated that in 1999, 8% of Brazil's soybean crop was GM varieties, even though Brazil had not yet officially approved their use. In addition, the government of Mexico recently announced that it had found genes engineered in corn among native maize varieties even though genetically modified corn seed has not been approved for sale in Mexico. More recent studies have not been able to confirm these findings.

Even though raw or minimally processed foods may be tested with sufficient accuracy, there is no consensus as to what method of testing should be utilized so that quantitative assessments between different laboratories may vary greatly. In addition, with increasing processing, the amount of false positives and false negatives increases due to denaturation of molecules being tested. Therefore, effective identity preservation is currently not achievable. To maintain accurate identity preservation, more sophisticated testing methods will have to be developed and the process must become standardized.

If a large enough market exists for non-GM foods, the food industry will respond. Consumers will likely have to pay more for GM-free foods, and truthful and nonmisleading labeling language will need to be defined by FDA. Organic foods could become the market channel for these GM-free foods. In any case, the next generation of GM crops might require identity preservation to capture value for farmers and processors.

Consumer education also plays a large role in this whole debate. The study sponsored by the Pew Initiative for Food and Biotechnology in January of 2001 found that the percentage of consumers who thought GM foods are safe increased from 29% to 48% after being informed that more than half of the products available at the grocery store include genetically modified foods. It is clear that agricultural biotechnology is now at a critical juncture, and ongoing educational efforts will need to be expanded both for opinion leaders and consumers. Media coverage will most surely grow on this subject as more products are introduced into the marketplace, but efforts are needed to ensure that the coverage is balanced. Ideally, the criteria most consumers will continue to use when selecting food will be taste, value, nutrition, safety and convenience instead of fear and emotion. If properly developed, agricultural biotechnology represents an opportunity for developing countries to realize tangible health and nutritional benefits with sustainable agriculture and new market niches. In fact, China and India are actively developing this technology, and their support for biotechnology may turn the tide in global public acceptance.

Agriculture is expected to feed an increasing population, forecasted to reach 8 billion by 2020, out of whom 6.7 billion will be in developing countries where the carrying capacity of agricultural lands will soon be reached given the current technology and the on-going environmental degradation (FAO, 1999). New technologies, if properly focused, offer a responsible way to enhance agricultural productivity at present and for the future. As suggested by the Nobel Peace Laureate, Norman Bourlag (1997), biotechnology and genetic plant engineering, complemented with conventional plant breeding, are needed to boost crop yields to feed the world.

Today, nearly 800 million people suffer from malnutrition; 1100 millions do not receive appropriate nutrition while subsisting below the levels of poverty (FAO, 1996a). In the case of the Latin American and the Caribbean (LAC) countries, in spite of the abundance of natural resources and continued investments in development, rural poverty and food insecurity have persistently affected more than 55 percent of the rural population with one third of the latter being below the extreme poverty line (income below a basic food basket). Furthermore, in six out of eleven countries for which there was information for several years, extreme poverty increased between the beginning of the eighties and the mid nineties. These elements establish the magnitude of the poverty and food insecurity to which research in agricultural production, comprising the application and complement of plant biotechnology, has to find a way to alleviate.

The food sector in the LAC Region is characterised, among other factors, by the growing dependence on food imports, primarily cereals, the decelerated growth of the agricultural production and poverty affecting wide sectors especially the rural population. Within this context, the increasing frequency of adverse environmental and climatic factors as drought and desertification, soil erosion, floods and hurricanes, and on-going changes in land use regarding other economic activities, are producing a significant reduction of the agricultural areas (James, 1996, 1997). Under these conditions, the use of sustainable technologies, traditional or modern, focusing on those with less input demands as well as emphasis on product quality and productivity are important factors for employment and income generation.

Fifteen years ago, plant biotechnology comprised only a few applications of tissue culture, recombinant DNA technology and monoclonal antibodies. Today, genetic transformation, and marker-aided selection and breeding are just a few examples of applications in crop improvement with profound implications for the LAC Region (Izquierdo, 1999). Plant biotechnology applications must respond to increasing demands in terms of food security, socio-economic development and promote the conservation, diversification and sustainable use of plant genetic resources as basic inputs for the future agriculture of the Region (FAO, 1999).

The scientific results obtained in the last two decades substantiate biotechnology as an option to increase food production. Some of the imperative reasons to use new biotechnology in an upward form reside in the fact that the benefits of the Green Revolution have already been surpassed and cannot guarantee the success of sustained food programs including their inability in terms of economic and environmental costs. Although the objective of the Green Revolution was to promote the production and the trade of cereals and other agricultural commodities, the majority of the developing countries have become net importers of basic cereal foods for daily human consumption and for livestock feeding (FAO, 1996b; Rosendal, 1992; Asiema, 1994).

The LAC agricultural sector needs a strong input of new knowledge and technologies to successfully confront the food insecurity challenge coupled with a realistic interpretation of the global environmental and economic situation (Commandeur, 1994; Vunne, 1995). It is in this complex food security context where biotechnology is called to play a growing role to contribute toward a safer world.

Food Security Concept

Food security is defined by FAO as the access by all people at all times to the food needed for a healthy and active life. The concept means the achievement of the food self-sufficiency, and guarantees that this condition will be sustained in the future. Food security also implies reaching a productive growth, compatible with the economic status of the producers and the preservation of the environment. The factors that determine the degree of food security, in any Region, country or zone in particular, are food availability, production stability and access to food by all members of the community (FAO, 1998).

Food availability is a factor directly related to the productivity of raw food supplies, as grains, root and tubers, oil and molasses. Variations in yield due to limitations of the crop genetic potential for a given environment and losses due to biotic, abiotic and post-harvest factors, greatly affects this issue.

Stability is related to the conditions of the productive process, including the sustained handling of natural resources, an appropriate policy for environmental protection, and the creation of an adequate rural infrastructure and political dynamics of rural development.

Access to food is a factor directly related to the prevailing socio-economic conditions in a particular zone or country. The lack of access to food is often exacerbated by the fact that producers try to keep higher prices by destroying large amounts of stocked foods or livestock. In addition, some countries that generate food surplus deny their access to others countries as a measure of political pressure. The causes and consequences of food insecurity and poverty are inextricably linked (FAO).

The *per capita* world food supply was equivalent to 2300 daily calories as an average in 1970. This figure included data from developed countries, whose average was of 3030 calories, as well as those from developing countries, where this indicator hardly reached to 1960 calories per capita, showing the enormous disproportion existing between both groups. 75% of the population of the developing countries, around 1600 million, corresponds to countries whose daily average per capita was nearly 1850 calories. The situation is even more precarious for the poorest sectors in the population due to the existing inequalities in each country (FAO, 1996c; Garett, 1997).

Chronic undernutrition and food insecurity are principally caused by: low productivity in agriculture, frequently caused in part by policy, institutional and technological constraints; high seasonal and year-to-year variability in food supplies, often the result of unreliable rainfall and insufficient water for crop and livestock production; lack of off-farm employment opportunities, contributing to low and uncertain incomes in urban and rural areas.

Opportunities and Scenarios for Plant Biotechnology Applications

Plant biotechnology represents one of a number of competing technological approaches to address any agronomic problem in a particular ecosystem but, as an example, a pest problem might equally be addressed through conventional plant breeding, through a transgenic approach, by the use of bio-pesticides, through an integrated crop management (ICM) approach or by any combination of these.

Andean agricultural scenarios in Bolivia, Chile, Peru, Ecuador and Venezuela, for instance, are depositories of important genetic resources that could constitute inputs to food security and sustainable development. Andean "orphan" food crops such as arracacha.(*Arracacia xanthorrhiza*), achira (*Canna edulis*), yacon (*Polymnia sonchifolia*), mashua (*Tropaeolum tuberosum)*, oca (*Oxalis tuberosa*), ulluco (*Ullucus tuberosus*), quinoa (*Chenopodium* quinoa), amaranto o kiwicha (*Amaranthus caudatus*), popping beans -"nuñas" (*Phaseolus vulgaris*), tarwi (*Lupinus mutabilis*), goldenberry-"capuli" (Physalis *peruviana*), cherimoya (Annona *cherimola*) and passion -fruit (*Passiflora sp.*) are highly under-utilised. The use of modern plant biotechnology for the conservation and sustainable agricultural use of those essential genetic resources needs to be strengthened. For this purpose, biotechnologies such as cell and tissue culture, molecular genome analysis, plant genetic transformation, molecular plant disease diagnosis and germplasm cryo-conservation coupled with plant breeding and physiological integrated crop management, can be successfully used to cope with genetic erosion. This can reinforce *ex-situ* collections and *in-situ* conservation, to upgrade the supply of improved and healthy seeds and planting materials for farmers and to integrate a new approach into the development programmes for food production and food security in mountain lands (Izquierdo and Roca, 1998). In the meantime, the population of the Andes suffer severely from the effects of ty, food insecurity, malnutrition and derived health problems.

The Amazon Basin is one of the richest areas in the world. The high plant and animal biodiversity gives this area a real potential to be sustainable in terms of food security, nutrition, and economy. However, the Amazon area in all of the countries where this very special humid forest exists, has security and malnutrition problems, as well as increasing poverty, specially in areas near the biggest cities of the Amazon (Izquierdo et al. 1999). The biggest problem in the Amazon has been the colonisation path. The countries in the Amazon have a very intense migration and have been penetrated with culture systems, especially animal production, that have caused a severe impact in this very fragile bio-system. The management of Amazon soils, their poor structure, low fertility and acidity, is one of the most important subjects to be studied and this is the principal concern of several R&D institutions in the Amazon countries. The gradual conversion of the on-going "extractivism" to sustainable fruit crop-agroforestry management, where all the productivity factors are considered, requires plant biotechnology applications for fruits, vegetables, medicinales, and forestry species including micropropagation and biotechnology-enhanced breeding.

Biotechnology is a powerful tool in agricultural development with great potential - both positive and negative. Coupled with other technologies, biotechnology can surely provide new solutions for some of the old problems hindering sustainable rural development and achievement of food security (FAO, 1999; Ortiz, 1998). Biotechnology also offers unique opportunities to solve environmental problems, some of which derive from unsustainable agricultural and industrial practices.

Plant biotechnology requires both human resources and infrastructure. The REDBIO Network has detected for this region technical constraints in terms of scarcity of human resources, limitations to technology development and use, infrastructure; institutional capacity and lack of information/data banks on plant biotechnology.

Food Security in the Latin American and Caribbean Region

According to the countries´ technical documents prepared for the Food World Summit, in the decade of the 70 the per capita food supply in the Region increased from 2360 to 2510 calories, while in the world it varied from 2300 to 2440 calories, compared with a global increase from 1960 to 2130 calories in the developing countries altogether (FAO, 1996a, b and c). This growing tendency was observed until the decade of the 80 when the per capita daily food supply went up to 2720 calories for the LAC Region, 2330 for the developing countries and 2580 for the world.

Unfortunately, these indexes did not represent a significant improvement in food security and remained practically unaltered (14% in 1980 and 15% in 1992 Considering the growth of the LAC population (279 million in 1970, 354 million in 1980, 443 million in 1992, 593 million estimated for 2010) malnutrition embraced 53 million people in 1970, 48 million in 1980 and 64 million in 1992. Today, it affects 15% of the population in Mexico, Central America and the Caribbean and 13% of the population in South America (FAO, 1996c and d) while the Region represents nearly 23% of the arable lands, 12% of the world cultivated areas, 46% of the forests, 31% of the world's fresh water reserves, and a rich ecosystem, with 50% of the plant and animal species of the biosphere representing an immense wealth in genetic resources.

Within the above context, food production growth should be sustained as a guarantee of food security (FAO, 1996a; Trigo, 1995) considering the differences among the LAC Region countries and internally among the different social sectors. This inequality is increasing and a society under such conditions cannot have food security. The levels of food security in the LAC Region have not increased and this situation should be reverted without delay.

To improve the levels of food security in countries with low income and food deficit, FAO has implemented the Special Program for Food Security (SPFS) as an assistance program to increase the food production and productivity on those countries less able to meet their food needs with imports. The objective is to reduce the present number of malnourished people in the world to one half, by the year 2015. Although in many of the LAC countries the demographic pressure is less than in other developing regions due to lower birth rate and better educational and health facilities (James, 1996; FAO, 1996; Garett, 1997; Trigo, 1995), a strong revitalisation of the agricultural sector is needed to secure food and rural income.

Agriculture is a fundamental sector for the establishment of an appropriate food production system. It has a marked multiplier effect in the general economic growth, generating employment and involving, besides its own activities, important economic sectors like transport, storage, processing and trade. The sector demands additional services beyond those already mentioned. Consequently, appropriate agricultural development can generate increased food availability, improving social development and providing better access to food.

Institutional Support

The development of biotechnology in the Region should be analyzed considering the real possibilities of the Region, country or among its different regions. Three categories of well-defined aspects: infrastructure, education and institutional support should be considered (Izquierdo, 1995). Infrastructure implies the establishment and assimilation of novel technologies and the implementation of material conditions according to the specific research project. This is a factor of difficult solution for the less developed countries since it implies strong investment (primary capital). Biotechnology is an emergent technological sector with the consequent economic risks. Only a strong governmental policy directed to promote this type of institutions until application and commercial use of results, will determine their gradual implementation and success for middle size projects or programmes.

The regional situation is affected by the growing activity of the transnational biotechnology companies that introduce complete technological packages with results ready for commercial exploitation. The relationship of these companies with the national biotechnology groups or programmes should contribute to the scientific development in the Region, but "clear cut" linking has to be established. Upgrading of research infrastructure, quantity and quality of trained human resources as well as appropriate norms for intellectual property rights would facilitate the implementation and financing of joint collaborative biotechnology research and development projects.

Institutional support is vital to maintain the available infrastructure, to modernise it and to maintain the qualified scientists and technicians formed during the implementation of the projects. The widespread concept on the role of institutional support limiting it to material aspect, is wrong. The institutions have the duty to assist the government in the implementation of a correct scientific policy and to define the objectives in the short, medium and long terms and, in consequence, to adequately use the available resources. They should have a scientific policy with enough flexibility to decentralise activities and functions and to allow for the creativity of the research groups according to national priorities.

Perspectives for Biotechnology Research on Food Security Projects in the Region

The success of projects and activities is limited by the present basic knowledge of the different research groups on plant biotechnology. This factor determines the feasibility of achieving results in a certain term and determines the research procedure and related risks.

A second factor is closely linked with the financing possibilities that depend on the priorities stipulated by the supporting sources. Frequently, laboratories and institutions are particularly attracted by those projects whose end products can be rapidly marketed at a good price, such as the ornamental and fruit-bearing plants, and those that allow an amplification of infrastructure and training possibilities, such as the "winter" multiplication of GMP´seeds in joint contracts with transnational companies.

The third factor refers to the international relationships of the different groups, searching for possibilities of international co-operation through multilateral or bilateral projects. Often the biotechnology groups look for a complementation of their profiles to present international research projects according to issues that are high-priority for the agencies or companies that support international co-operation, although that priority is not necessarily such within the national and regional context. The private sector is potentially interested in applications of biotechnology but it is necessary to carefully study necessities and priorities for proposals that can offer solutions and services related to food security and sustainable production.

Worldwide, the proportion of official financial support to agricultural research has decreased from 20% of the total investment in research in 1980 to 14% in 1990. This decreasing tendency, although smaller, continued in the late nineties. In spite of that, in the case of the LAC countries, it represents 90% of the total investment in agricultural research. In this aspect, biotechnology is still a second order address. However, the necessity of the producers for healthy and high genetic quality plant material, diagnosis services and certification and production of GMP seeds as well as norms related to biosafety, have increased as a consequence of the new policies in the regional economy and the efforts to look for new economic niches in the international market.

To achieve this objective the role that the public and private institutions should play to create or strengthen National Programmes to guide efforts toward the real necessities of production and food security, is important. This does not exclude the activities of advanced research laboratories or groups whose infrastructure, qualified personnel and connections with projects and programmes of developed countries allow them to develop this type of research.

Although the Region is characterised by low development in plant biotechnology, the total number of researchers under the national research systems is approximately 11000. At the present time, more than 1800 scientists, of which 1000 are postgraduates, are working in biotechnology projects within laboratories placed mainly in the universities, international research centers, and public and private institutions (REDBIO database,1999). However, due to the dispersion of the groups, faulty links with the productive sector and the absence of national and regional programs to organize and promote scientific activities, the coefficient of their real impact is considerably limited.

National Programs should identify objectives and priorities in each country, in order to promote regional collaboration, find financial support and strengthen the transfer of appropriate biotechnology and biosafety norms to the producers by promoting the application of valid results. Institutional support is reflected in all stages of scientific development, not only in research but also in the technological transfer to local producers. National institutions should promote the use of biotechnological advances for rural communities and the knowledge stemming from positive experiences to the whole system of agricultural-food production. Institutional support has been fickle in our Region and should be revitalised.

Public Knowledge and Attitudes

Although products derived from agricultural biotechnology are now common in the U.S. food supply, the topic does not appear to be an important issue to most U.S. consumers, and awareness remains moderate according to polls. Trends in consumer awareness and opinions about biotechnology and genetically engineered foods remain a shifting and somewhat controversial area. Polls designed to address similar issues at different times have found different results. Furthermore, such polls are sensitive to the wording and order of the questions. Perhaps the most telling information extracted from consumer polls are the trends over time, i.e., data obtained by asking the same questions after an intervening period of time. Notwithstanding the limitations of the technique, consumer polls remain an important tool in assessing public knowledge and opinion.

Evidently, most Americans are unaware of the extent to which crops derived from biotechnology have entered the marketplace. When consumers were asked how much processed food has genetically modified (GM) ingredients, only 14% responded with the correct answer (over 50% of processed foods have GM ingredients). Similarly, when asked whether or not they have eaten GM foods, 62% said no, they had not, while only 19% said that they had. With more than 60% of processed foods containing GM ingredients, it is highly probable that most people have eaten them at some time. Even though consumer awareness of biotechnology is moderate, knowledge levels tend to be quite low.

Nevertheless, more people are becoming aware of the issue. Between 1992 and 1996 only 34% of those asked said that they had heard either "a lot" or "something" about biotechnology. By 2000–2001 this number increased to 47%. Despite the controversy over the safety of eating GM foods, most American consumers seem to trust the food supply. When asked their willingness to buy GM potatoes or tomatoes protected from insect damage and requiring fewer pesticides, 74% in 1995, and 70% in January of 2001 indicated they were willing. It appears that the safety of eating foods derived from biotechnology is not on most people's minds. When asked an open-ended question regarding what they were most concerned about when it comes to food safety, the majority of consumers stated that they were worried about food handling, bacterial contamination, or pesticide residue issues. Very few people (2%) replied that they were concerned about biotechnology. In fact, when asked, "During the past few months, have you done *anything* or taken *any* action because of *any* concerns you may have about genetically modified foods?" only 5% of the respondents said "yes".

There is still a great deal of confusion concerning the basic science of genetics and biotechnology. In a survey conducted in 1998, 45% of participants disagreed and one tenth agreed with the statement, "Ordinary tomatoes do not contain genes, while genetically modified ones do." In addition, when presented with the statement, "By eating a genetically modified fruit, a person's genes could also be changed" only 61% disagreed and 9% agreed. Educating the public on these matters might prove to be a challenge as well. When asked of their trust in biotechnology information sources, 41% stated that they trusted the American Medical Association, 32% said they trusted the Food and Drug Administration, and 4% stated that they trusted activist groups to provide them with truthful information about biotechnology.

Based on perceived consumer concerns, Gerber, Heinz and Frito Lay have made the decision to avoid ingredients that were derived from GM crops. In fact, JR Simplot Co, a supplier of potatoes to fast-food chains, says virtually all it's customers don't want GM potatoes. Nevertheless, most food companies are not avoiding GM ingredients for domestic production, and by one estimate 60% of processed foods contain at least one GM ingredient (e.g., GM soy or corn flour).

Consumer interest is but one of many factors that influences the use of these ingredients in the production, processing, manufacture and distribution of foods and food products. Industry is concerned over the capability and costs to deliver identity-preserved ingredients, the adequacy of testing methodologies for GM constituents, uncertainties in forecasting supply needs, and increased costs due to shortages of available non-GM ingredients. Safety does not appear to be a significant concern because producers and manufacturers rely on the regulatory system in place to ensure the safety of GM products.

Natural Movements of Genes Between Species

Natural movement of genes between species, often called horizontal gene transfer or lateral gene transfer, can occur because of gene transfer mediated by natural agent This natural gene movement between species has been widely detected during genetic investigation of various natural Mobile genetic elements, such as transposons, and retrotransposons that naturally transfer to new locations in a genome, and often move to new species host over an evolutionary time scale. There are many types of natural mobile DNAs, and they have been detected abundantly in food crops such as rice.

These various mobile genes play a major role in dynamic changes to chromosomes during evolution and have often been given whimsical names, such as Mariner,

Hobo, Trans-Siberian Express (Transib), Osmar, Helitron, Sleeping Princess, MITE and MULE, to emphasize their mobile and transient behavior.

Such genetically mobile DNA contitutute a major fraction of the DNA of many plants, and the natural dynamic changes to crop plant chromosomes caused by this natural transgenic DNA mimics many of the features of plant genetic engineering currently pursued in the laboratory, such as using transposons as a genetic tool, and molecular cloning. *See also* transposon, retrotransposon, integron, provirus, endogenous retrovirus, heterosis, Gene duplication and exon shuffling by helitron-like transposons generate intraspecies diversity in maize.

There is new scientific literature about natural transgenic events in plants, through movement of natural mobile DNAs called MULEs between rice and Setaria millet. It is becoming clear that natural rearrangements of DNA and generation of transgenes play a pervasive role in natural evolution. Importantly many, if not most, flowering plants evolved by transgenesis - that is, the creation of natural interspecies hybrids in which chromosome sets from different plant species were added together. There is also the long and rich history of transgenic varieties in traditional breeding.

Deliberate Creation of Transgenic Plants During Breeding

Production of transgenic plants in wide-crosses by plant breeders has been a vital aspect of conventional plant breeding for a century or so. Without it, security of our food supply against losses caused by crop pests such as rusts and mildews would be severely compromised. The first historically recorded interpecies transgenic cereal hybrid was actually between wheat and rye (Wilson, 1876).

Last century, the introduction of alien germplasm into common foods was repeatedly achieved by traditional crop breeders by artificially overcoming fertility barriers. Novel genetic rearrangements of plant chromosomes, such as insertion of large blocks of rye (Secale) genes into wheat chromosomes ('translocations'), has also been exploited widely for many decades.

By the late 1930s with the introduction of colchicine, perennial grasses were being hybridized with wheat with the aim of transferring disease resistance and perenniality into annual crops, and large-scale practical use of hybrids was well established, leading on to development of Triticosecale and other new transgenic cereal crops. In 1985 Plant Genetic Systems (Ghent, Belgium), founded by Marc Van Montagu and Jeff Schell, was the first company to develop genetically engineered (tobacco) plants with insect tolerance by expressing genes encoding for insecticidal proteins from Bacillus thuringiensis (Bt).

Conserving Genetic Resources for Agriculture: Counting the Cost

As improved crop varieties developed by scientific breeding spread throughout the world in the latter half of the 20th century, the risk of excessive reliance by farmers and breeders on a narrowing genetic base was dramatized by the infestation and vulnerability of U.S. hybrid corn with cytoplasm male sterility to southern corn leaf blight. Events like this spurred worldwide efforts to greatly expand the amount of agricultural biodiversity conserved in genebanks. More recently, microarray and other modern biotechnologies that provide new and less costly ways of screening crop samples for useful traits have increased the value of conserved genetic resources and focused worldwide attention on access to and use rights of traditional crop varieties, or landraces, stored *in situ* (place of origin) or in *ex situ* genebanks worldwide. The 11 genebanks maintained by the research centers of the Consultative Group on International Agricultural Research (CGIAR) conserve more than 666,000 accessions (plant or seed samples) of crops grown mainly by poor people, staple food crops grown worldwide, and tree species used in agroforestry systems. This collection constitutes a sizable share—perhaps 30 percent or more—of the unique entries in genebank collections worldwide.

Conservation of this valuable germplasm should have a very long-term, if not perpetual, perspective. But funding for this long-term conservation service is currently provided on a precarious, year-by-year basis. This mismatch between the generally short-term nature of the financial support and the long-term nature and intent of the effort could threaten the security and future availability of this genetic material. A plan to judiciously match the duration of the funding commitments to the duration of the conservation commitments was unveiled at the World Food Summit in Rome in June 2002 and further elaborated at the World Summit on Sustainable Development in Johannesburg in August 2002. It involves an effort to tap private and public sources of support to establish a Global Conservation Trust (GCT) fund designed to sustain the long-term conservation and use of agricultural germplasm held in *ex situ* genebanks. But just how costly is it to conserve genetic resources in genebanks and maintain their viability and sample sizes in perpetuity? In this brief we estimate the costs of conserving specific crop species in *ex situ* genebanks in perpetuity, including the costs of maintaining healthy and viable seeds and other plant breeding material (collectively called "germplasm") stored in the field or in vitro.

We also show how these estimates change in response to variations among crops, conservation protocols, and institutional arrangements. The present value of these in-perpetuity costs indicates the necessary size of an endowment or trust fund that would furnish an income stream sufficient to underwrite long-term conservation efforts, thus keeping this valuable resource available for use in maintaining biodiversity and supporting plant breeding for the foreseeable future.

NOTABLE EVENTS IN THE HISTORY OF BIOTECHNOLOGY

- 8000 BC – Collecting of seeds for replanting. Evidence that Mesopotamian people used selective breeding (artificial selection) practices to improve livestock.
- 6000 BC – Brewing beer, fermenting wine, baking bread with help of yeast.
- 4000 BC – Chinese made yogurt and cheese with lactic-acid-producing bacteria.
- 1500 AD – Plant collecting around the world.
- 1590 AD – The microscope is invented by Zacharias Janssen.
- 1675 AD – Microorganisms discovered (using first microscope).
- 1856 AD – Gregor Mendel discovered the laws of inheritance.
- 1862 AD – Pasteur discover the bacterial origin of fermentation.
- 1919 AD – Karl Ereky, a Hungarian agricultural engineer, first used the word biotechnology.
- 1928 AD – Alexander Fleming noticed that a certain mould could stop the development of bacterias.
- 1953 AD – James D. Watson and Francis Crick describe the structure of Deoxyribonucleic acid, DNA.
- 1972 AD – The DNA composition of chimpanzees and gorillas is discovered to be 99% similar to that of humans.
- 1975 AD – Method for producing monoclonal antibody developed by Kohler and Milstein.
- 1980 AD –
  - Modern biotech is characterized by recombinant DNA technology. The prokaryote model, *E. coli*, is used to produce insulin and other medicine, in human form. (About 5% of diabetics are allergic to animal insulins available before).
  - A viable brewing yeast strain Saccharomyces cerevisiae 1026 acts as a modifier of the microflora in the rumen of cows and digestive tract of horses).
  - AD The United States Supreme Court 447 U.S. 303 (1980) rules in favor of microbiologist Ananda Chakrabarty in the case of a USPTO request for a first patent granted to a genetically modified living organism (GMO) in history.

- 1984 AD – Nutrigenomics as applied science in animal nutrition.
- 1994 AD – FDA approves of the first GM food from Calgene: "Flavr Savr" tomato.
- 1997 AD – British scientists from the Roslin Institute report cloning a sheep called Dolly using DNA from two adult sheep cells. Ian Wilmut led the team that cloned Dolly.
- 2000 AD – Completion of a, "rough draft," of the genome in the Human Genome Project.
- 2002 AD – Researchers sequence the DNA of rice, the main food source for two-thirds of the world's population. Rice is the first crop to have its genome decoded.
- 2003 AD – GloFish, the first biotech pet, hits the North American market. Specially bred to detect water pollutants, the fish glows red under black light thanks to the addition of a natural bioluminescence gene.

REFERENCES

Bud, Robert. "Biotechnology in the Twentieth Century." *Social Studies of Science* 21.3 (1991), 415-457.

Bud, Robert. "History of 'biotechnology.'" *Nature* 337 (1989), 10.

Bud, Robert. *The Uses of Life: A History of Biotechnology*. London: Cambridge University Press, 1993.

Bunders, J.; Haverkort, W.; Hiemstra, W. "Biotechnology: Building on Farmer's Knowledge." 1996, Macmillan Education, Ltd.

Colwell, Rita R. "Fulfilling the promise of biotechnology." Biotechnology Advances 20 (2002), 215-228.

D. Cameron, "Stop the Cloning", Technology Review, 23 May 2002'. Also available from http://www.techreview.com/. [hereafter "Cameron"]

Diaz E (editor). (2008). *Microbial Biodegradation: Genomics and Molecular Biology*, 1st ed., Caister Academic Press.

Dronamraju, Krishna R. *Biological and Social Issues in Biotechnology Sharing*. Brookfield: Ashgate Publishing Company, 1998.

Eugenics is the study of methods of improving genetic qualities through selective breeding

Feldbaum, Carl. "Some History Should Be Repeated." *Science* 295 (2002), 975.

Fiechter, Armin, volume editor. *History of Modern Biotechnology I*. New York: Springer-Verlag Berlin Heidelberg, 2000.

Gerstein, M. "Bioinformatics Introduction." *Yale University*. Retrieved on May 8, 2007.

Grace, Eric S. *Biotechnology Unzipped: Promises and Realities*. Washington, D.C.: Joseph Henry Press, 1997.

Krimsky, Sheldon. *Biotechnics and Society: The Rise of Industrial Genetics*. New York: Praeger Publishers, 1991.

Martins VAP et al (2008). "Genomic Insights into Oil Biodegradation in Marine Systems", *Microbial Biodegradation: Genomics and Molecular Biology*. Caister Academic Press.

Springham, D.; Springham, G.; Moses, V.; Cape, R.E. "Biotechnology: The Science and the Business." Published 1999, Taylor & Francis. p. 1.

Thackray, Arnold, ed. *Private Science: Biotechnology and the Rise of the Molecular Sciences*. Philadelphia: University of Pennsylvania Press, 1998.

U.S. Department of Energy Human Genome Program, supra note 6.

W. Bains, Genetic Engineering For Almost Everybody: What Does It Do? What Will It Do? (London: Penguin Books, 1987), 99.

2

Biotechnology and Genetic Resource: Uses, Research Areas and Regulations

BIOTECHNOLOGY AND GENETIC RESOURCE

But the new genetic modification (GM) technologies that many expect will help the world meet its food needs—not only through quantity, but nutritional quality as well raise critical issues for international trade, including this key question: What will happen if pressure from consumers and environmentalists in the developed world leads to a new generation of trade restrictions, or to the segmentation of GM-food product markets, as appears to be happening in Europe and Japan? An answer to this question requires a brief look at agricultural trade and involves both legal and economic analysis.

Agriculture and International Trade

Currently, a large share of agricultural production is consumed in the producing countries. This is true despite major grain and oilseed exports from countries such as Argentina, Australia, Canada, and the United States, and even after accounting for major export crops such as coffee, tea, cocoa, and sugar. IFPRI and others, however, forecast a growing role for international agricultural trade in the 21st century.

There is likely to be increasing specialization in agricultural production, with more exports from countries that specialize in particular types of agriculture. Many developing countries may well hold a comparative advantage in producing high-value, labor-intensive specialty crops and horticulture, while land-abundant countries may be better at producing bulk goods such as wheat, maize, and soybeans. Research indicates that it is neither efficient nor environmentally sound for developing countries to seek food security by becoming self-sufficient in the production of food crops, particularly when such production involves inefficient, unsustainable methods on fragile lands.

GM technologies may facilitate increased specialization, while also boosting local food production and improving food security through the development of plant varieties specifically tailored to particular agroecological environments. Although the technologies have the potential to affect both traded and nontraded products, most applications to date have involved highly traded agricultural commodities.

To benefit from increases in agricultural productivity, developing countries have an enormous interest in being able to market their goods in developed countries. The world agricultural trading system is still dominated by developed countries with protected markets and domestic subsidy programs that ultimately distort international markets and potentially increase price volatility, to the detriment of developing countries.

Major goals of developing countries in the new round of World Trade Organization (WTO) trade talks should include opening markets in developed countries for their agricultural exports, including high-value, labor-intensive commodities, and reducing or preferably eliminating trade-distorting domestic policies in developed countries—especially export subsidies and price supports.

While these goals appear desirable, the picture is complicated by the possible impact of consumer and environmental concerns, particularly within developed countries, on the development of biotechnology. To consumers in high-income countries, the price-reduction benefits from biotechnology seem minor, while the unknown dangers are magnified by lack of information and mistrust in the ability of their governments to regulate the safety of the food supply.

A ban on GM products in developed countries, based on domestic consumer and environmental concerns, not only would affect market access but could also make it more difficult for developing countries to gain financial support from industrialized nations to conduct research and build human capital for biotechnology activities. Another possibility is that consumer and environmental concerns could spill over into developing countries and block or slow the development of biotechnology in those countries.

International Legal Issues

Any attempt to limit trade in GM products must be compatible with existing international legal agreements. There are only a few agreements (including environmental treaties) setting out the WTO legal framework regarding trade in GM products. These include the Sanitary and Phytosanitary (SPS) Agreement and the Agreement on Technical Barriers to Trade (TBT) of the WTO as well as a multilateral environmental agreement, the Convention on Biological Diversity, and particularly its Cartagena Protocol on Biosafety.

The question is what role these legal agreements may play in either keeping open or closing the opportunities offered by GM products. The international system is clearly under stress in this area, with growing tensions between the need for fairness in international trade and the need to respond to domestic concerns about food and environmental safety.

The Sanitary and Phytosanitary Agreement, which concerns food safety and animal and plant health, says that WTO members have "the right to take sanitary and phytosanitary measures necessary for the protection of human, animal or plant life or health." But those measures must be applied "only to the extent necessary to protect human, animal or plant life or health" and must be "based on scientific principles." The agreement also states that WTO members must "ensure that their SPS measures do not arbitrarily or unjustifiably discriminate between Members where identical or similar conditions prevail, including between their own territory and that of other Members" and, furthermore, that those measures "shall not be applied in a manner which would constitute a disguised restriction on international trade." In addition, the agreement suggests the use of international standards when possible.

The goal of all these regulations phrased in legal language is to allow countries to maintain standards of food safety but to prevent them from doing so in a way that unfairly discriminates against foreign suppliers.

The difficulty with GM products is that there are as yet no international food safety standards that really apply to them. The Codex Alimentarius defines international standards of food safety, but it does not yet specifically address GM products. Although the countries participating in the Codex are currently discussing adequate standards for GM products, a possible agreement is still some years away.

In the absence of agreed-upon international standards, some countries invoke the "precautionary principle" that allows them to set standards provisionally where relevant scientific evidence is lacking, although they are supposed to do the necessary research within a reasonable period of time. Other countries argue that the precautionary principle is being abused in order to protect less-efficient domestic producers from foreign competition. Again, the challenge lies in adequately addressing both safety concerns and fairness in trade. Currently, a review of available scientific evidence indicates that GM foods have not been found to be unsafe—a double negative that highlights the difficulties of balancing consumer concerns, science, and international law. Proponents of GM products correctly argue that research has shown no health risks, while opponents argue that such research is not enough to prove that there are no such risks.

The basic issue continues to be market uncertainty about how consumers, mostly in developed countries, will react to GM foods. Regardless of the science, if consumers decide they do not want to consume GM goods, markets will adjust to satisfy their demands. If these negative reactions persist, markets will adjust to different scenarios of prohibition, market segmentation, and product differentiation. These market adjustments in developed countries will have an impact on developing countries.

The Economics of GM Trade

What will happen if consumers in developed countries refuse to consume GM commodities? Can world markets adjust to a complete segmentation of the markets for GM and non-GM commodities? Will developing countries still benefit from these new technologies if world markets are completely segmented and if, in addition, some developed countries refuse to adopt the new technologies at all? To provide tentative answers to these questions, IFPRI has undertaken research jointly with the Danish Institute of Agriculture, Forestry, and Fisheries Economics. Using multicountry models of world trade focused on agriculture, the research analyzes the price, production, and trade consequences of changing consumer preferences regarding the use of GM organisms in food production.

In the world model, the two primary GM crops, soybeans and maize, are specified as either GM or non-GM. This GM and non-GM split is maintained throughout the entire processing chain: GM livestock and GM food processing industries use only GM intermediate inputs; likewise, non-GM livestock and non-GM food-processing industries use only non-GM intermediate inputs. The underlying assumptions in the model are that developing countries will adopt the new technologies, to varying degrees, and that countries such as the United States will continue to use them, while Europe and Japan will not adopt them and will restrict their demand for such goods. The issue is which countries, if any, would benefit from the new technologies, to varying degrees, given the growing segmentation of the markets.

The empirical results show that global markets are able to adjust to this segregation in the sense that non-GM exports are diverted to the GM-intolerant regions, while GM exports are diverted to the indifferent regions. Price differentials are significant but tempered by commodity arbitrage. In particular, in certain GM- favorable regions, the prices of the non-GM varieties also decline because of the high degree of substitutability between the GM and non-GM varieties in domestic use and increased production of non-GM varieties to supply GM-intolerant consumers. The market results are analogous to what one would expect from increased consumer preferences in developed countries for organic foods. Such foods are more expensive to produce and command higher prices in the market. There is a gap between prices for organic and other foods, which ultimately reflects cost differences in their production and distribution. Similarly, price differentials between GM and non-GM commodities will reflect their different costs of production and distribution, with consumers who are indifferent benefiting from access to cheaper goods they find to be equivalent to non-GM goods and producers benefiting from the higher productivity of GM crops.

An important finding of this empirical analysis is ,that the developing countries are also responsive to GM preference changes and redirect their trade flows among partners accordingly. Furthermore, given the existing bilateral trade patterns for these particular crops, the price wedges that arise in the developing countries mainly reflect productivity differences, not preference changes in the developed world. Overall, the regions most receptive to the productivity- enhancing technology gain most, including developing countries that adopt the new technologies.

The development of GM technology appears to hold great promise, with the potential to complement other, more traditional research methods as the new driving force for sustained agricultural productivity growth in the 21st century. Such agricultural productivity growth is crucial if the world is to produce enough food to provide for what is likely to be a stable but large world population in this century. At this point, the many problems and concerns surrounding the new GM technologies do not seem insurmountable, just very difficult.

A world with an adequate supply of food is clearly more desirable than a Malthusian world in which food is scarce, food prices are high and rising, and people are in conflict over scarcity. Providing an adequate aggregate food supply will not eliminate malnutrition and hunger, however, now or in the future. To do that requires much more. To achieve food security for the entire world population, countries must work to reduce poverty and achieve a more equitable distribution of income—tasks that technology alone can only support, not achieve.

This suviary was formerly published in IFPRI Annual Report 2000–2001

Interlectual Properties and Developing Countries: Freedom to Operate in Agricultural Biotechnology

In agricultural biotechnology, the key technologies protected as intellectual property are highly concentrated in the hands of a small number of large, multinational corporations based in North America and Western Europe ("the North"). Although many developing countries ("the South") lack the capacity to adopt these technologies, a system of international and national agricultural research centers has used them to make genetic improvements benefiting the vast majority of poor consumers. Concern is arising in the worldwide agricultural research community that the very intellectual property rights (IPRs) that have been associated with the surge of private research in biotechnology now threaten to block access to new developments to public and nonprofit researchers. This concern about current developing-country access to essential intellectual property is exaggerated and largely misdirected. The relationship between IPRs and agricultural research in developing countries is poorly understood.

International and national agricultural research centers currently have far greater *freedom to operate*—the ability to practice or use an innovation—in agricultural research on food crops for the developing world than is commonly perceived.

The Misperception of IPRs

Even in developed countries, private sector agricultural research efforts concentrate primarily on a small number of crops with high commercial value. For the vast number of other crops, public and nonprofit institutions are the principal source of genetic innovation in the foreseeable future. In developed countries these institutions increasingly find their access to essential innovative inputs uncertain, unduly expensive, or at times blocked altogether (Wright 1998; Lindner 1999).

Given the minor role of the crops involved, this problem is a source of aggravation and inefficiency in the North but is in no way a serious threat to the well-being of consumers. Understandably, the international research and donor communities fear that the problems of access to intellectual property (IP) experienced in the North constitute a serious threat to the supply of food and fiber to the poor in the South. Many of the world's poor rely for sustenance on crops such as rice, beans, and cassava, which are largely beyond the focus of the private research sector and have modest commercial prospects due to low income elasticities. When major multinational corporations made some well-publicized "donations" of intellectual property to developing countries for certain noncommercial crops, they not only highlighted the usefulness of these technologies, but also reinforced the impression of a general lack of access to modern technological opportunities for these crops.

The Consultative Group on International Agricultural Research (CGIAR) and other international and local agricultural research organizations are still supporting and conducting agricultural research and development (R&D) geared toward poor farmers and consumers, as they did during the Green Revolution. The research budgets of many of these agencies, however, are now dwarfed by those of the major corporations in the field. Major donors have encouraged the CGIAR and other international and local agricultural research organizations to negotiate with major corporations to gain access to technologies for use in agricultural research conducted in or for developing- country economies. A survey shows fairly widespread use of protected IP by CGIAR centers, in many cases without formal authorization from the patentees (Cohen et al. 1998). While confirming the extent of international researchers' use of biotechnologies, this study showed researchers to be confused about relevant IPRs and created a sense of urgency about the regularization of licensing or other IPR transfer arrangements.

In fact, IPRs are based primarily on national laws. Public and nonprofit agricultural researchers generally have freedom to operate in regions where most modern technologies are unprotected by national IPR laws. Production in the South of a crop protected only in the North is both legal and moral per se (Barton and Strauss 2000; RAFI 2000). If, however, there is significant international trade in agricultural commodities and international transfer of the technologies used in their production, identifying valid IPR concerns becomes more complex. Thus, the spatial aspects of intellectual property are pivotal to freedom to operate in agricultural research.

The Rights to Research

The principal public policy rationale for protection of intellectual property is that it provides direct, socially beneficial incentives to innovate, while also facilitating further innovation by mandating public disclosure of the patented technology. When individuals or organizations know that legal protection will enable them to recoup their research investments, they have a stronger incentive to pursue such innovations. In the absence of protection, attempts to recoup investments or to profit commercially from an innovation may fail because of imitation. Knowing this, prospective innovators may underinvest in R&D or exploit their inventions in secret. In addition, by clarifying rights to new ideas, intellectual property protection helps to reduce the costs that would otherwise be required to determine ownership of rights.

An important but perhaps underappreciated aspect of most systems of IPRs is the requirement that inventors and researchers seeking these rights disclose the new knowledge they have obtained. As new ideas are disseminated through publication, licensing, or other means, this information stimulates further rounds of innovation and technological advances.

Inherent in intellectual protection is a tension between the goal of providing incentives for innovation and that of allowing innovators to build upon one another's work. The broader the monopoly rights conferred, the larger the potential threat to the freedom to operate. Owners of a technology may be unwilling to share or license it or willing only after costly negotiations, thus making it difficult for others to obtain essential tools for advancing their own research. Moreover, owners of technology may litigate against alleged infringers, so in practice, those who hope to use a protected technology must weigh the risk of litigation against the costs of obtaining licenses.

To further complicate matters, the modern methods used to develop new crop varieties depend on a wide range of component innovations, the rights to which may be held by many competing parties—be they patent rights or use rights assigned through commercial contracts or licenses. And the number of separate rights needed to produce a new innovation will only escalate as biotechnology patents become more prevalent. If ownership of these rights is diffuse and uncertain, it can be difficult or impossible for potential users to successfully negotiate with all of the relevant parties.

Yet agricultural researchers in many developing countries are freer than one might think to make use of innovations protected in the developed countries. This is because there is no such thing as an "international patent right." Patent or other rights awarded in, for example, the United States do not *a priori* confer property rights in the rest of the world. Patents and other IPRs are awarded by national governments, and the protection conferred by each national government applies only within that country. To obtain patent protection in several countries, innovators must apply for and gain rights in each. In countries where a technology is not subject to intellectual property protection, anyone is free to make, use, or sell whatever technology or knowledge is available for crops, irrespective of whether the crop is grown for subsistence or commercial use or the technology is protected elsewhere.

The extent of freedom to operate in developing countries is not well understood. For example, the recent vitamin A rice innovation (*Golden*rice) reportedly requires permission to practice more than 70 patent rights. The well-publicized donations by major corporations of their intellectual property relevant to vitamin A rice left a strong impression that they were relinquishing the exercise of large numbers of crucial patent rights in favor of the poor in developing countries. In fact, in some major rice-consuming countries, there are no valid relevant patents, and in most, there are very few. Similarly, the donations of virus-resistant technology for some noncommercial potato varieties in Mexico and for sweet potato in Africa apparently do not involve any patents relevant in the target countries. Finally, the Cohen et al. (1998) survey reported fairly widespread use of protected intellectual property by the centers of the CGIAR, in many cases without formal authorization from the patentees. But no distinction was drawn between patents valid in developed countries and those valid in the centers' host countries.

Though there is no international patent, international treaties and organizations do play an important role in IPR. They make it easier to extend protection to multiple countries and provide a uniform, minimal set of laws and standards that apply to all subscribing countries. Increasingly, innovators in developing countries are seeking IPRs in developed countries, and vice versa. Currently, however, in the fields of agriculture and agricultural biotechnology, the type and scope of protection varies greatly from country to country, especially between developed and developing countries. This variation makes it more difficult to assess whether there is freedom to operate on an international level.

How Production and Trade Patterns Affect IPRs

Understanding the production and trade status of crops relevant to developing countries is important not only in ascertaining the implications of IPRs, but also in assigning use rights by the private sector to public and nonprofit plant breeders. The willingness of owners of agricultural technology to cede use rights, or the minimum price at which they are willing to sell the rights to others, is shaped—among other things—by where crops are produced and traded.

Developing-world crop breeders have freedom to operate with respect to crops produced in developing countries unencumbered by local intellectual property protection of relevant inputs, processes, or products. Problems may arise, however, if those crops are subsequently exported in a form in which infringement is detectable to countries in which intellectual property protection is likely to prevail. In such cases it is the importer, not the breeder, who may be infringing on intellectual property. Binenbaum et al. (2000) studied production and trade data for 15 of the crops most important to research agencies operating in developing economies (soybeans, bananas, rice, coconuts, groundnuts, wheat, cassava, maize, beans, potatoes, chickpeas, sorghum, lentils, millet, and barley). The findings suggested the extent to which trade patterns are likely to raise IPR problems for agricultural research in developing countries:

(i) Exports from developing to developed countries of CGIAR crops are insignificant compared with total agricultural exports from developing countries, developed-country imports, or even domestic agricultural production, except for a few commodities and a few developing countries.

(ii) As a group the developing countries account for more than 90 percent of the world's production of rice, millet, cassava, sweet potatoes, yams, bananas, plantains, chickpeas, cowpeas, pigeon peas, groundnuts, and coconuts (and for quite a few of these crops they account for more than 98 percent of production). They also account for more than 65 per cent of the world's production of sorghum, beans, and lentils.

(iii) For the majority of CGIAR crops, output is never traded across international borders. Soybeans, coconuts, bananas, lentils, and beans are the only crops of the 15 studied for which more than 10 per cent of developing-country production is exported.

(iv) Just two crops (soybeans and bananas) account for 64 percent of developing-country crop exports to the developed countries, and just four countries (Argentina, Brazil, Costa Rica, and Ecuador) account for 42 percent of the South-North trade in these two crops. Adding exports of rice and coconuts amounts to 80 percent of the South-North trade total, with most of the rice coming from Thailand and coconuts from the Philippines.

(v) The principal destination for South-North trade in 9 of the top 10 developing-country crop exports (specifically soybeans, bananas, rice, coconuts, groundnuts, cassava, maize, beans, and potatoes) is Western Europe. Wheat is the only exception. To the extent that it is exported from developing countries, it is mainly shipped to North America and Japan. These exports are dwarfed, however, by wheat trade from North America to developing countries.

The trade data suggest that freedom-to-operate problems are most likely to arise in soybeans, bananas, and rice, but soybeans are not currently a major focus of public research by national or international agricultural research organizations working in or on behalf of the developing world. There is still substantial freedom to operate, however, for most crops of major significance for food security in poor countries. While freedom to operate in specific circumstances depends upon the claims of the IPR and its spatial pattern, crop production, and trade, IPRs over biotechnologies are mainly held in rich-country jurisdictions and are therefore primarily relevant to these jurisdictions. IPRs in the North affect farmers in the South if they export infringing products in detectable form to the North. South-North trade in food staples is limited overall, however, and involves only a few crops and developing countries in any significant way.

IPR-based limitations on export markets for food staples that embody technologies protected only in the North should not in general be considered an important impediment to the use of these technologies in such crops in the South. This does not mean that freedom to operate is no problem for developing-country research on exportoriented cash crops such as horticultural products, tropical beverages, or dessert bananas. The Binenbaum et al. study (2000) focused on the pre-dominant food crops of significance to poor people.

Focusing on More Urgent Problems Undue concern about the freedom to conduct research by or on behalf of developing countries is misdirecting policy and practical attention away from the main constraints currently facing researchers on food crops for the South. The real constraints are an increasingly serious lack of investment in developing-country research and a lack of local scientific skills to access the rapidly advancing stock of complex modern biotechnologies, whether they are protected by patents or not (Pardey and Beintema 2001). Biotechnology is challenging the adaptive capacity that has enabled poor countries to benefit from the advances in plant genetics and other relevant technologies in the past half-century, and lagging public resources are not being replaced by private-sector investments.

Failure to invest in the adaptive capacity needed to evaluate, access, and regulate the technologies being developed in the North is currently a far greater constraint than IPRs. The very confusion over this issue illustrates researchers' and decisionmakers' lack of capacity to handle questions relating to IPRs and freedom to operate in developing-country plant breeding. For the future, how the World Trade Organization's Agreement on Trade-related Aspects of Intellectual Property Rights (TRIPs Agreement) is implemented with respect to plant-breeding technology, domestically and in important export markets, is a crucial issue for developing-country policymakers. Where patenting of plant and other life forms is allowed, the patenting of key biotechnologies in the South will grow, threatening developing-country researchers' freedom to operate and freedom to trade in developing-country agricultural products, both South-North and South-South.

This issue ranks with implementation of farmers' rights as an important policy concern for plant breeders, farmers, and the food consumers of the South. But *domestic* freedom to operate is generally the relevant IPR issue; exports of food staples that dominate agriculture are not important growth drivers in most developing countries. Private corporations in the developed countries spent nearly US$11 billion on agricultural R&D in 1995 (in 1993 prices). By misunderstanding their present freedom to operate, breeders of food crops for the South threaten their ability to bargain effectively for access to the scientific outputs from OECD countries.

As institutional innovations bridging the private-public divide begin to emerge (Nottenburg et al. 2002), all parties need a clear picture of the present degrees of freedom regarding Southern agricultural R&D in order to strike effective deals when tapping Northern intellectual property on behalf of the world's poor, to know when such deals are not needed, and to recognize what is being surrendered in choosing patenting rather than plant breeders' rights in implementing the TRIPs Agreement.

Biotechnology, Production, and Agri-food Trade

Biotechnology is inextricably linked to international trade. The technology has been globally developed and is being applied to research programs in more than 30 countries around the world. Biotechnology has had the greatest effect on the most heavily traded agri-food commodities in the global trading system.

Although the first biotechnology-based agri-food product entered the market only in 1994, by 2001 more than 50 modifications involving 13 crops had been approved and produced on more than 52 million hectares in at least 14 countries. Commercial production of GM foods has been concentrated in canola, corn, cotton, and soybeans, which are extensively traded internationally. Perhaps most important, GM production has been concentrated in countries that are the traditional and dominant exporters of those crops (particularly Argentina, Canada, China, and the United States). Up to 88 percent of trade in some of the products with GM varieties comes from the key GM-adopting countries. For the most part, GM products have been marketed as commodities and mixed with batches of GM and non-GM products as they flow into the international market- place and then to many countries around the globe. Once in these markets, the commodities are extensively processed, and their components (edible oils, corn meals, soybean proteins, and so on) are fundamental ingredients in more than 70 percent of the processed foods available in most developed-country markets.

GM products appear to simply raise new concerns about access to international markets. Those few countries producing and exporting the products seek to be able to continue their business unimpeded. Yet GM varieties tend to exacerbate the debate about market access because almost all the biotechnology traits in commercial production—herbicide tolerance, insect resistance, and viral resistance—lower production costs or increase yields. Those countries adopting these technologies, which also tend to be traditional exporters, thereby increase their exportable surpluses, depressing world prices and making nonadopting importing producers less competitive. As a result, disadvantaged farmers may join with consumers in importing countries concerned about the safety of these products in calling for increased controls on these products.

Global Regulatory Response

Nine international bodies are currently vying to coordinate and regulate different aspects of food safety. These institutions fall into three types. Five are largely science-based organizations: the International Plant Protection Convention (IPPC), International Epizootics Organization (OIE), Codex Alimentarius (Codex), the Food and Agriculture Organization of the United Nations (FAO), and the World Health Organization (WHO). One, the World Trade Organization (WTO), is a trade-based organization. The three others have broader objectives such as environmental protection and other social or political goals: the Organisation of Economic Co-operation and Development (OECD), Regional Initiatives, and the Cartagena BioSafety Protocol (BSP). These organizations seek to develop standards for health, safety, and labeling for GM foods, establish testing procedures to ensure the standards are met, provide rules for allowable policies, and create systems to manage disputes.

Despite the substantial effort being undertaken, there is no common view on the goal of international regulation. While most agree that safety is the bottom line, few can agree on what that means, whose opinions should hold the most weight (scientists' or citizens'), or how to handle nonsafety issues such as social, economic, or ethical concerns. The FAO and WHO have a long history of multilateral efforts to promote food security and public health and have worked to develop a consensus about the implications of biotechnology for their areas of interest. Meanwhile, the IPPC and OIE are multilateral treaties that seek to protect plants and animals from the spread of pathogens through international trade, thereby providing much of the scientific consensus that underlies domestic food safety systems. Both institutions have their own nonbinding dispute avoidance and settlement systems, but their most important role in international trade is through the WTO Sanitary and Phytosanitary Agreement (SPS), which uses the IPPC and OIE standards as the basis for evaluating SPS disputes. National measures based on international standards from either of these institutions will generally not be open to challenge under the WTO dispute resolution process.

Furthermore, both the IPPC and OIE nominate experts for WTO SPS dispute panels and provide technical background information to the panels based on their standards. As such, they can have far-reaching economic and political consequences on food trade.

The Codex, under the joint FAO/WHO Food Standards Program, provides a similar service related to processed foods. The Codex develops international food standards, which identify the product and its essential composition and quality factors, identify additives and potential contaminants, set hygiene requirements, provide labeling requirements, and establish the scientific procedures used to sample and analyze the product. Each standard normally takes six or more years to develop. Determination of the safety of the food product is based on scientific risk analysis and toxicological studies. Once a Codex standard is adopted, member countries are encouraged to incorporate it into any relevant domestic rules and legislation, but they may unilaterally impose more stringent food safety regulations for consumer protection, provided the different standards are scientifically justifiable. Codex plays an important role in agri-food trade because its standards, guidelines, and recommendations, like the IPPC and OIE provisions, are acknowledged in the SPS and Technical Barriers to Trade Agreements during consideration of trade disputes. There has been an eight-year process to develop a Codex standard for products of biotechnology, but consensus eludes the negotiators.

The OECD, composed of 29 industrial democracies, has actively assisted in harmonizing international regulatory requirements, standards, and policies related to biotechnology since 1985. The OECD has undertaken a number of projects to make regulatory processes more transparent and efficient, to facilitate trade in the products derived through biotechnology, and to provide information exchange and dialogue with non-OECD countries. A number of bilateral or multilateral regional initiatives have played an increasingly important role in regulating trade in goods and services. These institutions help create the consensus necessary to establish international rules, given that many food safety concerns in trade are bilateral and the knowledge base to develop standards resides in a few countries only.

The Trans-Atlantic Economic Partnership (TEP) between the United States and the EU, for example, has undertaken talks in recent years to improve regulatory processes and scientific cooperation through mutual recognition of testing and approval procedures; progressive realignment or adoption of the same standards, regulatory requirements, and procedures; the adoption of internationally agreed upon standards; and dialogue between scientific and other expert advisers in standard-setting bodies and regulatory agencies. The EU has similar trade liberalization initiatives with Canada and Japan.

Since 1998 the Canadian Food Inspection Agency and the U.S. Department of Agriculture's Animal and Plant Health Inspection Service have also been studying and comparing the molecular genetic characterization of transgenic plants in search of ways to harmonize their regulatory review processes. Some agreement has already been achieved, although no formal binding bilateral agreement has yet been concluded. Meanwhile, Canada, the EU, and the United States all offer training and support for regulators in key import markets (usually developing countries) in an effort to "export" their regulatory models to other countries.

These bilateral processes could be an important way to resolve technically based trade disputes. Regional agreements, memoranda of understanding, mutual recognition agreements, formal dialogues, and joint research projects are mechanisms that can be used to decrease bilateral regulatory barriers to GM food trade. The WTO has become a focal point for examining and resolving trade disruptions related to GM foods.

Although there was a nonbinding agreement on technical barriers to trade in the Tokyo Round of the General Agreement on Tariffs and Trade, the 1995 SPS agreement for the first time extended the newly formalized and binding dispute settlement system to cover trade concerns related to sanitary and phytosanitary rules and technical barriers to trade. The WTO agreement permits national "standards or regulations for the classification, grading or marketing of commodities in international trade" (Article XI) and the adoption or enforcement of measures necessary to protect human, animal, or plant life or health (Article XX(b)), but it sets some rules on when and how they may be used. Specifically, the SPS Agreement requires that measures

(1) do not discriminate between member states; (2) conform where possible to international standards developed by Codex, OIE, or IPPC; (3) be based on scientific principles and the completion of a risk assessment study; and (4) do not constitute a disguised restriction on international trade. Although the WTO is the main locus of dispute resolution for many countries, it has some limitations. As currently interpreted, the SPS Agreement allows regulations based on science but does not permit regulations that restrict trade based on nonscience concerns such as consumer preference, animal welfare, or nonmeasurable environmental risks.

The Cartagena Biosafety Protocol is one effort to provide a more comprehensive international structure to ensure the protection of biodiversity and to facilitate consideration of nonscientific concerns in food trade. Although the Cartagena Protocol, concluded in Montreal in January 2000, is primarily designed to provide rules facilitating advance informed agreement (AIA) for first-time transboundary movements of living GM organisms intended for environmental release, it also provides for labeling (but not AIA) of GM elements in commodity shipments destined for the food chain. Countries can use this transparency to decide whether to import those commodities, but the current interpretation is that import bans must still be consistent with the WTO principles already noted. It is perhaps too early to make a confident evaluation of the protocol.

The only conclusion one can derive from this survey of international institutions is that no one institution, and perhaps not even the entire array of institutions, is likely to yield an early resolution to concerns about diverging national policies and regulations concerning GM foods.

The adoption of biotechnology and the introduction of GM foods into the international marketplace has exacerbated an already difficult area of trade policy. As biotechnology increases productive capacity in various products, it also increases the need to trade. But diverging national regulations are increasingly impeding trade in these products. This situation has begun to create production and trade distortions, which will build over time. Overcoming these distortions is made more difficult by the fact that the recent WTO agreement on agriculture is not yet fully implemented, and many of the issues left to handle are highly contentious. There is little goodwill in the policy community that can be directed to resolving the growing trade irritants caused by GM foods. As a result, a messy trade world is likely to continue. The private sector may find it needs to change how it introduces and markets the new products of biotechnology in order to maintain market access.

Local Regulatory Response

A number of factors have made this issue hard to handle. Uncertainty about the food and environmental safety of new GM foods has led to different responses in different markets. Those markets lacking domestic regulators that command the confidence of consumers have tended to act in a "precautionary" way, either reviewing the products more slowly or imposing tem- porary bans on the introduction of the new products. This is a sharp break from the international food safety system that evolved over the past 100 years, where importers tended to accept the food and environmental safety judgments of regulators from those countries developing and exporting the products. One result of this "renationalization" of agri-food safety regulation is that national systems have tended to diverge. Canada, Japan, Mexico, and the United States, among others, generally make similar rulings and have approved most of the new GM products for production and consumption. Regulators in Australia, the European Union (EU), and New Zealand, in contrast, have postponed approvals in recent years, reflecting the concerns of their citizens. Another 20 or so countries have developed domestic regulatory systems consistent with one or other of these approaches.

The diverging domestic systems are most evident when one looks at the labeling systems being proposed or developed in various countries (Phillips and McNeill 2001). So far more than 26 countries have either adopted provisions or announced plans for rules to help the market develop and deliver labeled products. At one extreme, Argentina, Canada, Hong Kong, and the United States have adopted a voluntary labeling strategy that will likely allow labels for either GM or GM-free products, with only 1–5 percent tolerances for comingling. At the other extreme, 22 countries and the EU have adopted or announced plans to implement mandatory labeling systems. As of June 2002, only a handful of these countries had revealed the full structure of the labeling rules they intend to pursue, and only Australia, China, Japan, New Zealand, South Korea, and the United Kingdom have formally implemented labeling systems. A number of other countries have proposed mandatory labeling (for example, Brazil, Czech Republic, Hungary, Indonesia, Poland, Russia, South Africa, and Thailand), but there is little available evidence that these countries have developed domestic systems to manage such regulations or, for that matter, any firm indication of when their systems might be operational.

The key concern about the diverging domestic regulatory systems is that production and trade are shifting. Key GM adopters, especially Canada and the United States, are abandoning or losing key markets and diverting their exports to new markets. U.S. exports of corn to the EU have fallen by 70 percent in recent years, U.S. exports of soybeans to the EU have dropped by 48 percent, and Canadian exports of canola to the EU have dropped 96 percent. Meanwhile, the EU has developed new GM-free sources of soybeans from Brazil and canola from Australia, both markets that have not yet approved GM varieties for those crops. So far these changing trade flows have not significantly affected producer returns—trade has simply been reallocated between adopting and nonadopting countries—but over time such policies have the potential to seriously distort trade flows and offset many of the benefits of recently negotiated international trade agreements for these products.

Most of the rest of the countries in the world do not have any domestic regulatory capacity and are seeking guidance and help from international institutions.

Assessing Others People's Technology

Public and private nonprofit institutions worldwide engaged in agricultural research and biotechnology are increasingly active participants in intellectual property transactions, interacting with the for-profit sector and even spawning private entities of their own. Notably absent from the group of nonprofit institutes seeking patent protection are the 16 centers of the Consultative Group on International Agriculture Research (CGIAR). Located primarily in developing countries, only a few centers have obtained patent protection for their inventions.

Nonprofit research institutions are not in the business of selling products to consumers. If they are to realize a return on their investment, they must sell rights to their technologies to commercial entities or other research institutions rather than make them freely available. A nonprofit entity may, for example, exclusively license technology to a commercial partner, license the technology itself nonexclusively, or use the technology as the foundation for a spin-off company.

For all the benefits that nonprofit institutions receive from intellectual property, these same institutes are notorious for using other people's patented technologies without permission. A review of the intellectual property policies of several large universities in the United States with active licensing offices reveals that none discusses the need to obtain permission to use patented methods and materials, and only one provides guidelines on copying material that is copyright protected (Nottenburg, Pardey, and Wright 2002). In contrast, for-profit entities—especially in biotechnology—are not only generally more cognizant of intellectual property rights and rules, but also proactive in obtaining licenses, options for licenses, or collaborations that will assure their "freedom to operate," that is, their ability to practice or use an innovation.

Nonprofit research organizations need to develop and implement policies regarding use of other people's technologies. With a special emphasis on agricultural biotechnology, this brief discusses policies of intellectual property protection, de jure (by right) and de facto research exemptions, and the ways that research at nonprofit institutes fits with, and is at odds with, these policies and exemptions. We also present an overview of the steps necessary to abide by others' intellectual property rights (IPRs) and show how most non-profits are ill equipped to undertake such measures.

Finally, we present strategies for pursuing different options to obtain rights to use other people's technologies.

Protecting Intellectual Property

The major forms of legal protection available for agricultural biotechnology are patents, plant breeders' rights (known in the United States as Plant Variety Protection Certificates), trademarks, trade secrets, and contracts. Third-party trademarks and trade secrets, however, have relatively little impact on nonprofit institutions and so will not be discussed here. Protecting and controlling the use of intellectual property can also be achieved by technical means, like hybridization of crops such as corn and rice and genetic use restriction technologies (GURTs). These methods have the greatest impact on farmers by rendering the seed unsuitable for replanting or suppressing the expression of certain introduced traits in saved seed. They are dealt with in detail by UNEP/CBD/SBSTTA (1999).

A web of proprietary claims now envelops the transfer and use of patented biotechnologies, thereby limiting the freedom to operate of public and private agencies alike. Biotechnologies covered by these claims include (1) parent germplasm in the form of individual plant varieties; (2) germplasm constructs that include trait-specific genes; and (3) transformation technologies, such as a gene coding for a specific characteristic inserted into plant cells, selectable markers, and gene silencing or regulating technologies. Depending on the complexity of the transgenic product, dozens of identifiable proprietary claims can be involved in its development.

Patents

Patents protect inventions of tangible things and confer a legally enforceable right on their owners to exclude others from practicing the invention described and claimed in the document. These rights apply for a limited period of time, generally 20 years, and only in a specific legal jurisdiction.

A common misconception is that a patent awarded in one country confers rights in the rest of the world. This is not so; there is no such thing as an "international patent." Patents are awarded by national governments, and the protection conferred by a patent extends only to the national jurisdiction in which the patent is awarded. To protect an innovation in more than one country, a patent must be awarded in each. The cost of obtaining a patent varies from country to country, and the cost of obtaining protection in every important market can be substantial. In general, most inventions are patented in just one or a few countries, mainly the developed ones.

Plant Breeders' Rights

To be granted a plant breeders' right (PBR), an applicant must demonstrate that the variety is new, distinct from other varieties, and genetically uniform and stable through successive generations. The holders of a PBR have a legal monopoly over commercialization of their variety for a prescribed length of time. Generally, PBRs encompass the right to sell, reproduce, and import a new plant variety. PBRs in most jurisdictions contain a research exemption.

Forms of PBRs consistent with the International Union for the Protection of New Varieties of Plants (UPOV) now exist in most developed countries. Developing countries are adopting either UPOV standards or other forms of plant variety protection to comply with the requirement of the Trade-Related Aspects of Intellectual Property Agreement (TRIPs) to grant a so-called *sui generis* form of protection to plant varieties. By December 2001, 50 countries (including, most recently, Bolivia, Brazil, China, and Kenya) had enacted PBR legislation.

Permission Issues

The nature of the patent right allows the patent holder to exclude others from making, using, selling, offering for sale, or importing the patented invention. To encourage basic research, countries have sought to facilitate access, either through a statutory exemption, a common law exemption, or compulsory licensing.

In the United States researchers generally assume that patent law does not apply to their basic research. Academic researchers are often shocked to discover that, except for some limited statutory exemptions, there is no general research exemption for using other people's patented technologies. Courts generally have ruled that using another's invention for research or experimental use is an infringement. Research at a university or other nonprofit institution, even if performed without any profit motive, would constitute an infringement.

The U.S. Congress has enacted only a few narrow exemptions. Yet there does appear to be a de facto exemption in the United States. Even absent a legal research exemption, it is unlikely that nonprofit institutions have more than a minor risk of infringement exposure, especially in cases where the nature of the research is clearly noncommercial. The number of patent suits filed against nonprofit organizations in U.S. District Courts is extremely small.

Commercially Oriented Research

The risk of infringement liability may be higher when commercially oriented research or services are involved. In these cases the unauthorized user may receive a letter requesting that the activity cease and desist, an offer for a commercial license, or notice of an infringement action.

An important trend, however, is that the line between nonprofit and commercially oriented research is becoming blurred. An increasing amount of research is performed as part of private-public sector alliances. Substantial private sector funding also supports research conducted by government agencies and public universities in many developed countries, and in some developing ones as well. Public policies have encouraged this. In the United States, for example, the Bayh-Dole Act of 1980 mandated that the U.S. government cede ownership of intellectual property emanating from government-sponsored research tothe recipient institutions. Under the auspices of Cooperative Research and Development Agreements (CRADAs), specifically designed to speed the commercialization of federally developed technology, the government and its collaborating partner may share patents and patent licenses, allow one partner to retain exclusive rights to a patent, or assign licensing rights to facilitate licensing to third-party users.

What is "Free Access"?

Given the risk of using other people's patented technologies, some in the nonprofit research world may want express permission to use the technologies. Permission may be obtained in a variety of ways, but the recipient should be vigilant in identifying the hidden costs of access. Sometimes agreements widely characterized as onerous are actually far less restrictive than apparently "free" deals and traditional consulting arrangements between private firms and individual academics.

Determining Freedom to Operate

As nonprofit research becomes more commercially oriented, the risk of serious consequences for infringement may well increase. As risk increases, the need to scrutinize the intellectual property landscape and the freedom to operate will become more pressing. There are various reasons why determining freedom to operate can be a daunting task, especially for the nonlegal professional.

— A freedom-to-operate analysis is, by design, a snap- shot of the current patent situation; however, patenting and disclosing inventions is a dynamic process. A review of emerging publications is integral to such analyses given the continuous stream of patents and applications being published.

— The challenges inherent in an ever-changing landscape are further complicated by the difficulty of determining which entity will triumph, and with what claims.

A patent's claims—not its text—define the parameters of the patent right conferred on the patentee. Hence, to delineate the extent of the right, a potential user must interpret these claims. In the United States claim construction is a matter of law and centers on an objective test of what a person of ordinary skill in the art at the time of the invention would have understood the claim to mean. Infringement is determined by examining whether the alleged infringing product or method falls within the scope of the claims.

The development of any product in biotechnology requires a multitude of technologies and reagents. This is especially true in agricultural biotechnology, where the delivery system includes germplasm. Typical reagents include vectors for transformation of plants, components of vectors, elite plant varieties, and the like. In the case of *Golden* Rice TM), an analysis estimated that 70 patented technologies were used during research and development (R&D). This analysis illustrates the complexity of intellectual property in agricultural biotechnology.

Several databases with differing amounts of information are available on the Internet; some are available by paid subscription and some are free. For non-legal professionals, a problem common to all the existing databases is the interface, which caters to individuals with a substantial knowledge base concerning intellectual property. Furthermore, with the exception of the database of the Center for the Application of Molecular Biology to International Agriculture (CAMBIA; see www.cambiaIP.org), none provides an explanation about patents, how to read a patent, or other information to assist the naïve user.

Options for Gaining Access to Other People's Technology

Various options are available for gaining access to proprietary technologies. Some of the more important ones are discussed here, mainly from the perspective of a nonprofit agency. This discussion emphasizes developing countries, although most of the issues are relevant in developed countries too.

Cross-Licensing

At CGIAR centers, licensing would have to be restricted to property other than landraces and other plant varieties designated as "in trust" material (under a 1994 agreement with the Food and Agriculture Organization of the United Nations) that must be made available to the world at large. Through a material trust agreement (MTA), recipients of intrust material distributed by CGIAR centers agree not to seek intellectual property protection on that material, though they may seek protection for derivatives.

Despite these severe constraints, candidates for cross-licensing have already been identified. The near-isogenic lines of rice germplasm potentially useful in plant breeding and developed at the International Rice Research Institute are examples of plant breeding that might be licensed via an MTA or other contractual agreement. Fischer and Barton (1999) proposed a model MTA in which a CGIAR center would offer such material to another institution at no cost in exchange for access to information about subsequent discoveries and zero-cost nonexclusive research licenses to CGIAR centers and agricultural research agencies operating in developing countries. If this example leads to successful cross-licensing, it is likely to be the exception that proves the rule. The number and value of intellectual property resources held by most public agencies operating for developing countries are often overstated, which puts them in a relatively weak negotiating position.

Research-Only Licenses

A free research license that does not permit commercialization can make a research tool the "cuckoo's egg" of technology transfer. If the project succeeds, then the bargaining for permission to commercialize must begin. On the one hand, researchers have already incurred the sunk cost of all the research, placing them in a highly disadvantageous bargaining position. On the other hand, even in refusing to allow commercialization, the IPR holder gains valuable information about the technology and its downstream applications.

Market Segmentation Strategies

All CGIAR centers engaged in biological research are in developing countries. Patents are usually filed in, at most, a select group of countries. Indeed, until recently few developing countries allowed patents on life forms. To the extent that research agencies use technologies and cultivars that are not patented or otherwise protected where the agencies are located, they can and should legally proceed without obtaining permission from the IPR holder. Even after compliance with TRIPs, the breeding of new cultivars using prior cultivars protected in developed countries may be legal under the sui generis protection that is being adopted in many developing countries.

The new regime of the World Trade Organization (WTO) might facilitate a kind of indirect market segmentation, in which developing countries get the new technology for free, and proprietary claims are enforced in developed countries. Further, cultivars incorporating genes patented in developing countries may not be subject to effective intellectual property claims if those countries have neither the legal means nor the will to enforce them.

In the near term, research agencies in developing countries are likely to have considerable freedom to operate, if they operate judiciously. Because retroactive patenting is impossible, most of the technologies usable by the CGIAR and its developing-country partners over the next half-decade or so are likely to be unencumbered by relevant intellectual property rights. Mistakes, however, could result in catastrophic legal liability. To reliably implement a strategy of obtaining intellectual property only where necessary, those who make research commitments must have access to adequate information and to expert legal counsel. Such access does not exist for most developing- country researchers and research institutions.

A promising initiative to provide intellectual property information services for developing-country organizations is being pursued by the Australian non-profit corporation CAMBIA. The aim is to develop interactive software that can help researchers identify prior patent claims and areas of freedom to operate and thus travel more safely through the international patent minefield. If adequately funded on a continuing basis, such an initiative could reduce the uncertainties about prior claims to useful biotechnology.

Markets for intellectual property can also be segregated on grounds other than geography. With technology licenses, common segmentation strategies include delineating fields of use, length of time, certain claims of a patent, limitations to specific uses of the technology, research use versus commercialization, or restrictions on third-party services. Another option is to charge license fees based on an ability to pay or expected profit streams.

Mergers or Joint Ventures

Mergers can be a way to avoid an expensive patent fight. In agricultural biotechnology, mergers are a prime private sector solution to minimize the private costs of transactions in intellectual property. Mergers and outright privatization of previously public research agencies are characteristic of public sector agricultural R&D reforms in countries such as the Netherlands and the United Kingdom. But much of this change seems to have been driven by policy reforms and public budget cuts, not by a consideration of intellectual property.

Joint ventures are often viewed as a more promising and flexible alternative. For example, Monsanto is marketing transgenic cotton in China in a joint venture with a provincial public seed-producing organization.

Cost-Free Licensing of Technologies

For many minor crops, private and public IPR holders might be persuaded to allow international agricultural research centers and public research agencies in developing countries to develop proprietary biotechnology for use by farmers without any direct compensation. This situation is more likely where there is obviously little risk to the significant commercial markets that are the focus of the IPR holders' hopes for profits. Such cases have already occurred in these noncom- mercial crops, including several under the auspices of the International Service for the Acquisition of Agri-Biotech Applications (ISAAA).

Direct Programmatic Research Support from the Private Sector

For-profit corporations might be persuaded to give more general support to collaboration with public research. Important examples of such support on the part of corporations with significant market power have already occurred. In the genomics field, a consortium of corporations has supported creation of a public database of genome markers in preference to par-taking in a competing private sector initiative. Such cases suggest that private firms might, on occasion, choose to support public or private research initiatives in areas complementary to their own endeavors.

In another case Monsanto donated technology for the transformation of corn by *Agrobacterium* to the University of California. As part of a divestiture of assets ordered by the U.S. Justice Department, Monsanto was persuaded to give this technology to the university, allowing the university to license access to the technology to third parties. The details of this case illustrate the important point that prospective recipients must exercise flexibility and initiative to take advantage of such opportunities.

Patent Pools

A patent pool is an aggregation of intellectual property rights that are cross-licensed and licensed to third parties. In the United States the two critical features of a patent pool are that (1) the pool integrates complementary patent rights, and (2) the resulting competitive benefits are likely to be outweighed by competitive harm posed by other aspects of the program. Thus, patents in the pool must be essential to practice the technology.

Such joint agreements are probably not feasible as a regular modus operandi for pooling agricultural biotechnologies on a one-by-one basis. A better option is to coordinate a joint commitment by the major biotechnology providers and public agencies (including the CGIAR) to provide royalty-free licenses on all IPRs in agreed terms of application. In negotiating and drafting any such agreement, attention should be paid to national antitrust laws. This type of negotiation is difficult and costly to all parties and requires high-quality legal advice.

Clearinghouse Mechanisms

An alternative means of lowering the cost of technology transactions in biotechnology is the creation of an Internet-based clearinghouse (Graff and Zilberman 2001). This clearinghouse could identify relevant intellectual property in specified technology endowments, its availability, and how it could be obtained. It could also establish prices or pricing indicators, facilitate negotiations, and offer mechanisms for arbitration of disputes and monitoring of compliance. An agricultural biotechnology intellectual property clearinghouse could bundle together sets of complementary patents from different patent holders into complete "biotechnology or agronomic systems" contracts. Through such strategies, it would be possible to create customized licenses that could greatly increase the use of inventors' technologies and make multipatent technology systems readily available and affordable to researchers.

Independent Development of Research Tools

A quite different approach is to sponsor the creation of substitutes for existing proprietary research paths. For example, CAMBIA seeks to generate new biotechnology tools for agriculture, unencumbered by restrictive property claims. These tools are in turn made available on an ability-to-pay basis. The licensing revenues are used to fund further research and to support transfer of the technologies to developing countries.

Pressure for Sharing of Technology

International research institutions, including the CGIAR and FAO, should continue to press for including the interests of international and developing-country nonprofit research collaborations in measures designed to address the interests of domestic research institutions in the leading countries.

One form of pressure is a boycott of companies demanding "unreasonable" terms for key enabling technologies. Making common cause with more powerful allies in applying pressure on IPR holders might help ensure that any concessions by IPR holders are extended to nonprofit international agricultural research and that intellectual property is disseminated to noncommercial markets.

Designing policies and operating procedures to ensure that public science has sufficient freedom to operate is becoming increasingly important in the developed and developing worlds. Freedom to operate will be crucial for public and nonprofit agencies intent on developing improved seed varieties and other technologies destined for commercial release, albeit in markets that may generate large social gains but are not necessarily privately profitable. Various options are available to improve the efficiency of public-private relationships—particularly options that could lower the transaction costs of tapping proprietary technologies to further public research. Paradoxically, for developing countries the short-run importance of freedom to operate has been exaggerated by well-publicized donations that generate inferences that the multinational life science oligopoly holds extensive portfolios of intellectual property that block further research in those countries. Ironically, in developed countries nonprofit researchers often believe themselves exempt from infringement suits. Worldwide, institutions need to better understand their rights and responsibilities regarding intellectual property.

As things stand now, intellectual property does not appear to be the binding constraint on science in developing countries, but it is becoming a constraint on nonprofit research in rich countries. The real problems facing many countries and agencies, especially in developing countries, are lack of local investment in science and limited experience and expertise in gaining access to, using, and regulating modern biotechnologies. Developed countries are not immune to these problems either. Also suffering are the agricultural biotechnology industries in developed countries like Australia and Canada, which have comparatively small investments in domestic R&D but are highly dependent on exports to countries that have strong intellectual property protection (such as the United States and European countries). Furthermore, the implementation of TRIPs as currently formulated will likely affect the freedom to operate in the next generation of biotechnologies. Guiding these changes in intellectual property regimes and responding creatively to the new environment are pressing challenges for those interested in the future of scientific research, including agricultural biotechnology.

USE OF BIOTECHNOLOGY IN DIFFERENT AREAS

Biotechnology techniques in modern times are currently being used in many areas such as food, agriculture, health care, forestry, the environment, minerals, oil and gas recovery and industrial processes to develop new products and processes, and to modify existing ones. Summarized below are some of the ways that modern biotechnology techniques have been applied in these areas.

Food and Agriculture

One of the most extensive applications of biotechnology has been in agriculture. Biotechnology techniques such as recombinant DNA techniques and mutagenesis have been used to develop plants with novel traits. These traits include herbicide tolerance and pest, insect and virus resistance. Biotechnology techniques have been used to produce biopesticides which are toxic to targeted plant pests. Also, experiments into genetic modification of aquatic organisms such as salmon, for such novel traits as enhanced growth, have been carried out using recombinant DNA techniques.

Health Care

To date, applications of biotechnology in health care have focused on fighting disease using the human body's own 'weapons'. Biotechnology medicines and therapies synthesize proteins, enzymes, antibodies and other substances which occur naturally in the human body, to fight infections and diseases. However, biotechnology also uses other living organisms, that is, plant and animal cells, viruses and yeasts to help produce human medicines. There are four main areas in health care in which biotechnology is currently being used: medicines, vaccines, diagnostics and gene therapy. Several biotechnology medicines have been developed and approved to treat such diseases as anaemia, growth deficiency in children, hemophilia, diabetes and others. Biotechnology companies have also developed therapeutic products for infectious agents such as HIV, hepatitis B and influenza. Biotechnology is responsible for new vaccines such as the hepatitis B vaccine which was approved in the U.S. to fight the hepatitis B virus. Biotechnology diagnostics have been used to detect a wide variety of diseases and genetic conditions. For example, gene therapy is used diagnostically to screen donated blood to protect the blood supply from HIV and hepatitis. Biotechnology diagnostics are also used in home pregnancy tests. In a home pregnancy test kit, a protein called a monoclonal antibody (MAb), binds to Human Chorionic Gonadotrophin (HCG), causing a colour change. HCG is present in a woman's urine only during pregnancy. In gene therapy, genes are used as 'drugs' to treat hereditary genetic disorders. Faulty or missing genes can be replaced to prevent the occurrence of a genetic disease. Current gene therapy is primarily experiment-based, with a few human clinical trials, such as for the treatment of cystic fibrosis, in the early stages.

Biotechnology and Environment

Biotechnology applications in the environment focus on using living organisms primarily to treat waste and prevent pollution. Examples of these applications include bio-filtration and bio-remediation. Bio-filtration refers to the use of microorganisms to remove complex pollutants from the air emissions and waste water discharges of many manufacturing processes. Bio-remediation refers to a number of processes that use living organisms to degrade toxic waste into harmless by-products such as water, carbon dioxide and other materials. Examples of bio-remediation processes include:

- *Bio-stimulation:* A technique which involves introducing nutrients to stimulate the growth of waste-eating microorganisms already present in the environment at a waste site.
- *Bio-augmentation:* A technique whereby microorganisms, not normally present in the ecosystem of a contaminated site, are added to the site to clean up pollutants. The microorganisms added may be natural or genetically modified. The organisms die when their food is used up.
- *Phytoremediation:* A technique which uses certain plants and fungi, planted at waste sites, to naturally reclaim the sites without using chemical treatment, incineration or land filling. These plants and fungi are used because they flourish by accumulating metals and other waste materials present in the soil.

Biotechnology and Forestry

To date, only a few biotechnology-derived products and processes used in forestry have been commercialized. Most biotechnology applications in forestry are still in the research and development stage. However, some commercial applications include the development of bacterial bio-pesticides such as *Bacillus thuringiensis* (Bt), as an alternative to chemical pesticides. Tissue culture technology is also being used to produce genetically modified seedlings for forest regeneration. For example, BCRI Inc., a biotechnology company in British Columbia, is using tissue culture to produce weevil-resistant Sitka spruce. Companies are also developing bio-fertilizers such as plant growth-promoting rhizobia bacterial fertilizers. In addition, genetic engineering has been used to develop traits of agronomic interest such as pest and disease resistance, within trees. The pulp and paper industry has used enzymes and microorganisms in their bio-bleaching and de-inking processes.

Biotechnology and Industrries

Industrial biotechnology (also known as white biotechnology) is the practice of using cells to generate industrially-useful products. *The Economist* speculated (as cited in the *Economist* article listed in the "References" section) industrial biotechnology might significantly impact the chemical industry. *The Economist* also suggested it might enable economies to become less dependent on fossil fuels. Diversa is an example of a company that specializes in industrial biotechnology. A significant problem in industrial biotechnology is waste production. A cell may be used to generate desirable carbon dioxide, other cells, and other molecules. It will use energy to accomplish its industrial purpose. Yet it will also use some energy to generate waste (like acetic acid) instead of the desired product or products. Decreasing waste production is a significant goal in industrial biotechnology. Metabolic engineering may help reach that goal.

Biotechnology has been applied in a variety of industrial processes in different ways, particularly in the use of biocatalysts in manufacturing processes. Biocatalysts are substances that initiate or modify the rate of a biological process and are generally consumed in the process. Some examples of industrial processes where biotechnology has been applied include chemicals, starch/grain processing, cleaning and textile industrial processes. In the chemicals industry, biotechnology has been used to produce commodity and specialty chemicals. It is also used in the starch and grain processing industries through the use of enzymes to turn starch into glucose and fructose. Corn and other grains can be converted to sweeteners such as high-fructose corn syrup and maltose syrup, using enzymes. Biotechnology applications have been used to produce ethanol from grain. Furthermore, biotechnology is used in the textile industry for the finishing of fabrics and garments. In the pulp and paper industry, bio-pulping is used to manufacture some products.

Related Industrial Technologies

This looks at how biotechnology contributes to new and enhanced industrial processes. It examines how biotechnology enables the use of renewable resources in industry, thereby increasing efficiency, decreasing pollution, and reducing energy usage and waste production during various types of industrial processing.

Biopolymers and Bioplastics

Many everyday products used by Canadians are made of plastic. Pens, computers, cars, food packaging, and clothing are all examples of products which contain plastic. Plastic is a material made up of one or more polymers. The main source of the chemicals needed to manufacture plastics are fossil fuels. These petrochemical plastics are very durable, but take a long time to biodegrade when disposed. Rising concern about the cost of fossil fuels, and their impact on the environment has resulted in a search for alternatives to petrochemical plastics, namely biopolymers and bioplastics.

Biopolymers and bioplastics go by many different names. They are often referred to as bio-based plastics and polymers, or as biodegradable plastics or polymers. They are defined as: (i) Biopolymers are polymers which are present in, or created by, living organisms. These include polymers from renewable resources that can be polymerized to create bioplastics and (ii) Bioplastics are plastics manufactured using biopolymers, and are biodegradable.

Biopolymers and bioplastics are not new products. Henry Ford developed a method of manufacturing plastic car parts from soybeans in the mid-1900s. However, World War II side-tracked the production of bioplastic cars. Today, bioplastics are gaining popularity once again as new manufacturing techniques developed through biotechnology are being applied to their production. There are two main types of biopolymers: those that come from living organisms; and, those which need to be polymerized but come from renewable resources. Both types are used in the production of bioplastics. These biopolymers are present in, or created by, living organisms. These include carbohydrates and proteins. These can be used in the production of plastic for commercial purposes.

Biotechnology in the Textiles Industry

Biotechnology has impacted the textiles industry through the development of more efficient and more environmentally friendly manufacturing processes, as well as through the design of improved textile materials. Some of biotechnology's key roles have involved the implementation, production, and modification of enzymes for the improvement of textile manufacturing processes. Biotechnology has also facilitated the production of novel and biodegradeable fibres from biomass feedstocks.

Enzyme Biotechnology in Textiles

Through biotechnology, enzymes are used to treat and modify fibres during textile manufacturing, processing, and in caring for the product afterwards. Some applications include:

- De-sizing of cotton – Untreated cotton threads can break easily when being woven into fabrics. To prevent this breakage, they are coated with a jelly-like substance through a process called sizing. However, after the threads have been woven into fabrics, the agents needed to further finish the material cannot adhere to the jelly-coated fabrics. Thus, the protective sizing agents must be removed by a process called de-sizing. Amylase enzymes are widely used in de-sizing, as they do not weaken or affect cotton fibres, nor do they harm the environment.
- Retting of flax – Flax plants are an important source of textile fibres. Useful flax fibres are separated from the plant's tough stems through a process called retting. Traditional retting methods consume large quantities of water and energy. Bacteria, which may be bred or genetically engineered to contain necessary enzymes, can be used to make this a more energy efficient process.
- Breakdown of hydrogen peroxide – When cotton is bleached, a chemical called hydrogen peroxide, which can react with other dyes, remains on the fabric. Catalase enzymes specifically break down hydrogen peroxide and may be used to remove this reactive chemical before further dyeing.
- Biostoning and Biopolishing – Instead of using abrasive tools like pumice stones to create a stonewashed effect or to remove surface fuzz, cellulase enzymes may be used to effectively stonewash and polish fabrics without abrasively damaging the fibres.
- Detergents – Enzymes allows detergents to effectively clean clothes and remove stains. They can remove certain stains, such as those made by grass and sweat, more effectively than enzyme-free detergents. Without enzymes, a lot of energy would be required to create the high temperatures and vigorous shaking needed to clean clothes effectively. Enzymes used in laundry detergents must be inexpensive, stable, and safe to use. Currently, only protease and amylase enzymes are incorporated into detergents. Lipase enzymes break down too easily in washing machines to be very useful in detergents. However, their stability is being studied and further developed through methods such as genetic screening and modification.

Synthetic fibres made from renewable sources of biomass are environmentally sustainable, and are becoming increasingly economically sustainable. Biodegradable synthetic polymers include novel fibres such as polyglycolic acid and polylactic acid, which are made from natural starting materials. Not all novel fibres are synthetic; they may also be naturally derived. Some natural biological fibres come from basic materials found in nature, including:

- Chitin – a type of sugar polymer found in crustaceans
- Collagen – a type of protein found in animal connective tissue
- Alginate – a type of sugar polymer found in certain bacteria

A prime example of a synthetic biomass fibre is Polylactic Acid (PLA), which is made by fermenting cornstarch or glucose into lactic acid, and then chemically transforming it into a polymer fibre. With properties similar to other synthetic fibres, PLA based materials are durable with a silky feel, and may be blended with wool or cotton.

PLA has potential applications in several areas, including the following:

- Textiles – clothing, fashions, upholstery
- Agriculture – plant mats, tree nets, soil erosion control products
- Sanitation – household wipes, diaper products
- Medicine – disposable garments, medical textiles

PLA minimizes environmental waste, as it may be fully biodegraded by microorganisms under appropriate conditions into carbon dioxide and water. Unlike the non-renewable petroleum resources used to make traditional synthetic fibres, the supply of renewable corn biomass needed to make PLA is expected to surpass demand in the anticipated future.

Biodegradable synthetic fibres and natural biological fibres may be used to make textiles for medical applications. Such textiles may be used in first aid, clinical, and hygienic practices. Some examples are described below:

Biochemicals

Biochemicals are chemicals produced from organic sources known as biomass. They are sometimes called bio-based chemicals, green chemicals, and plant-based chemicals. Biochemicals are chemicals produced from organic sources known as biomass. Biomass is any living matter that can be naturally and regularly replenished, including agricultural food and feed crop residues, aquatic plants, animal wastes and other waste materials. Biochemicals are sometimes called bio-based chemicals, green chemicals, and plant-based chemicals. Currently, the most popular source for biochemicals is corn. Chemicals like ethyl lactate, succinic acid, and polylactic acid are made from corn components, such as corn starch, and are used to make a variety of products, such as plastics, solvents, clothing fibres, paints, and food additives. Petroleum is the oil found underground in certain parts of the world and is the most used energy source in the world. It is used to make gasoline, chemicals, plastics and drugs. Because petroleum is limited in quantity, and cannot be made once the supply has been depleted from the earth, it is said to be non-renewable. Petroleum products are also not environmentally friendly – causing air pollution when burned, and filling up landfills when disposed of. Biochemical products made from biomass help solve many of the problems associated with petrochemical products. Environmental benefits include less pollution from crude oil extraction and processing, as well as better waste management since the materials are biodegradable.

The role of biotechnology in the creation of biochemicals is crucial. Through fermentation technology, biotechnology enables the processing of common crops and crop wastes, such as corn kernels and corn stalks, to make products such as plastics, textiles, and ethanol. Biotechnology also allows the creation of genetically modified bacteria that can convert plant sugars into an ingredient used in clothing, packaging, and plastics. New enzymes are also being developed for breaking down plant sugars. Biotechnology also helps scientists better understand the biology of bacteria and enzymes that can be used in bioprocessing, which may enable the development of cheaper methods of biochemical production. Currently, biochemical products are often expensive to produce and are not cost-effective enough to compete with traditional products.

Research in biochemicals continues to grow rapidly. Many large multinational companies are investing time and money into producing chemicals that can be economically viable and environmentally friendly. For example, many chemical companies are teaming up with biotechnology companies to develop products based on biochemicals. Cargill Dow currently makes polylactic acid from corn kernels but has plans to switch to cheaper feedstocks, such as corn stalks, wheat straw, rice hulls, and sawdust, so that biochemicals can be more competitive in the marketplace. Current research is also focussed on developing genetically modified microorganisms for use in specific chemical productions. Pollution due to the use of petroleum-based products may be eliminated by the use of biochemicals. Whereas petroleum-based products do not biodegrade and accumulate in waste landfills, biochemicals degrade completely and do not contribute to waste problems. As well, biochemicals are made from biomass; therefore, their production and use are sustainable.

Polymer Use(s)

Polylactic Acid and Polyglycolic Acid: Used in sutures, absorbable wound closure products, orthopaedic repair absorbable pins, and fixation devices, as well as in tissue engineering structures

Collagen: Uses in cell engineering structures, such as in artificial skin, or even as surgeon's thread

Synthetic biomass fibres may also be used in drug delivery systems, which are designed to release drugs at a specified rate for a specified time.

Enzymes in Industrial Applications

Enzymes are proteins that speed up biochemical reactions without being consumed or changed by the reaction. They are found throughout nature; in our bodies, in the environment, and in all living things. Without enzymes, life would not be possible. Enzymes speed up chemical and biochemical reactions, and this process is called catalysis. The substances upon which enzymes work when performing catalysis are known as substrates, and each enzyme will only fit and act upon a specific set of substrates.There are so many enzymes that it would be impossible to name them all. In fact, scientists have yet to discover many enzymes, or fully understand their structure and properties. On the other hand, many other enzymes have been successfully studied and applied to industrial and commercial uses.

In the absence of enzymes, chemical reactions occur only when molecules collide while in proper alignment with each other. Because molecules are bumping into each other randomly, chemical reactions are essentially due to chance events. This sometimes results in reactions that occur very slowly, or reactions that do not occur at all.Enzymes act like tiny molecular machines to ensure that molecules come into contact with each other and react. Like a key fitting into a lock, chemical molecules fit into pocket-like structures located on an enzyme. These pockets hold the molecules in a position that will allow them to react with each other, ensuring that they are close enough together and aligned properly for a reaction to occur.

In this way, enzymes speed up reactions. The enzymes are not changed themselves by the reaction. When the reaction is complete, enzymes release the product(s) and are ready to bring together more molecules and catalyse more reactions. Biotechnology and EnzymesBiotechnology can provide an unlimited and pure source of enzymes as an alternative to the harsh chemicals traditionally used in industry for accelerating chemical reactions. Enzymes are found in naturally occurring microorganisms, such as bacteria, fungi, and yeast, all of which may or may not be genetically modified.

Large quantities of enzymes are often needed for industrial use, so these microorganisms are multiplied through a process called fermentation. When enzymes or the microorganisms used to create them are no longer needed, they are destroyed through exposure to heat or safe organic/inorganic materials.

Enzyme Biotechnology in the Pulp and Paper Industry

Enzymes have been used in the pulp and paper industry to soften wood fibres, improve drainage, and present alternatives to chemical bleaching.

- Biopulping - Paper is made from cellulose fibres, which must be separated from a tough wood fibre called lignin. The step by step process used to separate cellulose from lignin and other wood components is known as pulping. It is a time and energy consuming process, involving the mechanical processing of wood or the treatment of wood with harsh chemicals. In biopulping, cellulase and xylanase enzymes made by lignin-degrading fungi are used to pre-treat wood and break down the lignin fibres. Removing lignin prior to further wood pulping saves time and energy, and decreases the quantities of chemicals used.
- Draining - Enzymes can also improve water drainage during wood pulping, a process that often slows down paper production. When fine lignin fibres are degraded by enzymes, less water is absorbed, thereby reducing drainage times, lessening the energy required to dry the paper, and producing a cleaner water runoff.
- Bleach Boosting - Lignin fibres that remain in wood pulp are coloured and must be bleached, usually by harsh chlorine compounds under high pressures. As an alternative, enzymes may be used to remove fine surface fibres, thereby reducing the bleaching process or eliminating it altogether.

Enzyme Biotechnology and Textiles

Enzymes are used to treat and modify fibres, particularly during textile processing and in caring for textiles afterwards. For example, enzymes called catalases are used to treat cotton fibres and prepare them for the dyeing processes. Some bacterial enzymes are used to separate the tough stem of the flax plant from the flax fibres used in textiles. By degrading surface fibres, many enzymes, including some cellulases and xylanases, are used to finish fabrics, give jeans a stonewashed effect, or help in the tanning of leathers. A recombinant enzyme called laccase, made by certain fungi, may also be used to treat fabrics and even catalyse the synthesis of some synthetic fibres. There are even enzymes in regular laundry detergent to help break down dirt, clean clothes more effectively, and prevent the dulling of fabric colours.

Enzymes are frequently used in laundry detergents. Without them, very high temperatures and mechanical shaking would be required to effectively clean clothes and other textiles.

Enzyme Biotechnology and Biofuels

Enzymes may be used to help produce fuels from renewable sources of biomass.

Such enzymes include cellulases, which convert cellulose fibres from feedstocks like corn into sugars. These sugars are subsequently fermented into ethanol by microorganisms. A new process called Simultaneous Saccharification and Fermentation has greatly improved ethanol production efficiency. In this new process, cellulase enzymes and fermentation microorganisms are combined in a single reaction mixture to produce ethanol in one step, rather than producing sugars from cellulose and then fermenting them into ethanol separately.

Enzymes are a sustainable alternative to the use of harsh chemicals in industry. Because enzymes work under moderate conditions, such as warm temperatures and neutral pH, they reduce energy consumption by eliminating the need to maintain extreme environments, as required by many chemically catalysed reactions. Reducing energy consumption leads to decreased greenhouse gas emissions by power stations. Enzymes also reduce water consumption and chemical waste production during manufacturing processes. Because enzymes react specifically and minimize the production of by-products, they offer minimal risk to humans, wildlife, and the environment. Enzymes are both economically and environmentally feasible because they are safely inactivated and create little or no waste; rather than being discarded, end-product enzymatic material may be treated and used as fertilizer for farmers' crops. Bottom of Form

Microbial Enhanced Oil Recovery

Microbial Enhanced Oil Recovery (MEOR) is the use of microorganisms to retrieve additional oil from existing wells, thereby enhancing the petroleum production of an oil reservoir. In this technique, selected natural microorganisms are introduced into oil wells to produce harmless by-products, such as slippery natural substances or gases, all of which help propel oil out of the well. Because these processes help to mobilize the oil and facilitate oil flow, they allow a greater amount to be recovered from the well.

Biotechnology has been used in the mining industry in such processes as bio-leaching and metal bio-remediation and recovery. Bio-leaching is the use of bacteria to extract valuable metals. Metal bio-remediation and recovery refers to the use of biologically created enzymes in the degreasing process to recover metals of value.

Biotechnology and Energy

To date, biotechnology applications in the energy industry include the production of cleaner coal and petroleum by removing sulfur and thus reducing the environmental contaminants released during combustion. Biotechnology has also been used widely to produce bio fuels such as bio-ethanol and bio-diesel. Bio fuels are alcohols, ethers, esters and other organic chemicals made from biomass such as herbaceous and woody plants, agricultural and forestry residues and industrial waste. These fuels can be used for electricity and as fuels for transportation.

Biotechnology and Human Applications

When treating problems that arise from genetic disorder, one solution is gene therapy. A genetic disorder is a situation where some genes are missing or faulty. When this happens, genes may be expressed in unfavorable ways or not at all, and this generally leads to further complications. The idea of gene therapy is that a non-pathogenic virus or other delivery system can be used to insert a piece of DNA—a good copy of the gene—into cells of the living individual. The modified cells would divide as normal and each division would produce cells that express the desired trait. The result would be that he/she would then have the ability to express the trait that was previously absent at least partially. This form of genetic engineering could help alleviate many problems, such as diabetes, cystic fibrosis, or other genetic diseases. The potential of genetic engineering to cure medical conditions opens the question of exactly what such a condition is. Some advocates see aging and death as medical conditions and engineering problems to be solved. They see human genetic engineering as a key tool in this. The difference between cure and enhancement from their perspective is merely one of degree. Theoretically genetic engineering could be used to drastically change people's genomes which could enable people to regrow limbs, the spine, the brain. It could also be used to make people stronger, faster, smarter, or to increase the capacity of the lungs, among other things. If a gene exists in nature, it could be brought over to a human cell. Others feel that there is an important distinction between using genetic technologies to treat those who are suffering and to make those who are already healthy superior to others. There is widespread agreement that germline engineering should not currently be allowed for either therapeutic and enhancement applications, as evidenced by a recent report by the American Association for the Advancement of Science. Though theory and speculation suggest that genetic engineering could be used to make people stronger, faster, smarter, or to increase lung capacity, the AAAS report finds that there is little evidence that this currently can be done safely or without unethical human experiments.

"Transgenic organism" is now the preferred term for genetically modified organisms with extra-genome (foreign genetic) information, as opposed to "genetically engineered" or "genetically modified" organisms (which may refer to changes made within the genome such as amplification or deletion of genes). The first Genetically Engineered drug was human insulin approved by the USA's FDA in 1982. Another early application of GE was to create human growth hormone as replacement for a drug that was previously extracted from human cadavers. In 1986 the FDA approved the first genetically engineered vaccine for humans, for hepatitis B. Since these early uses of the technology in medicine the use of the GE has expanded to supply many drugs and vaccines. One of the best known applications of genetic engineering is that of the creation of genetically modified organisms (GMOs). There are potentially momentous biotechnological applications of GM, for example oral vaccines produced naturally in fruit, at very low cost. A radical ambition of some groups is human enhancement via genetics, eventually by molecular engineering. DNA sequencing is a technique which is used to identify each base in DNA. Although the costs of DNA sequencing has dropped dramatically, the NIH estimates it costs at least $10 million to sequence 3 billion base pairs - the size of the whole human genome. Although there has been a tremendous revolution in the biological sciences in the past twenty years, there is still a great deal that remains to be discovered. The completion of the sequencing of the human genome, as well as the genomes of most agriculturally and scientifically important plants and animals, has increased the possibilities of genetic research immeasurably. Expedient and inexpensive access to comprehensive genetic data has become a reality, with billions of sequenced nucleotides already online and annotated. Now that the rapid sequencing of arbitrarily large genomes has become a simple, if not trivial affair, a much greater challenge will be elucidating function of the extraordinarily complex web of interacting proteins, dubbed the proteome, that constitutes and powers all living things. Genetic engineering has become the gold standard in protein research, and major research progress has been made using a wide variety of techniques, including:

- Loss of function, such as in a knockout experiment, in which an organism is engineered to lack the activity of one or more genes. This allows the experimenter to analyze the defects caused by this mutation, and can be considerably useful in unearthing the function of a gene. It is used especially frequently in developmental biology. A knockout experiment involves the creation and manipulation of a DNA construct in vitro, which, in a simple knockout, consists of a copy of the desired gene which has been slightly altered such as to cripple its function. The construct is then taken up by embryonic stem cells, where the engineered copy of the gene replaces the organism's own gene. These stem cells are injected into blastocysts, which are implanted into surrogate mothers. Another method, useful in organisms such as Drosophila (fruit fly), is to induce mutations in a large population and then screen the progeny for the desired mutation. A similar process can be used in both plants and prokaryotes.
- Gain of function experiments, the logical counterpart of knockouts. These are sometimes performed in conjunction with knockout experiments to more finely establish the function of the desired gene. The process is much the same as that in knockout engineering, except that the construct

is designed to increase the function of the gene, usually by providing extra copies of the gene or attracting more frequent transcription.

- 'Tracking' experiments, which seek to gain information about the localisation and interaction of the desired protein. One way to do this is to replace the wild-type gene with a 'fusion' gene, which is a juxtaposition of the wild-type gene with a reporting element such as Green Fluorescent Protein (GFP) that will allow easy visualisation of the products of the genetic modification. While this is a useful technique, the manipulation can destroy the function of the gene, creating secondary effects and possibly calling into question the results of the experiment. More sophisticated techniques are now in development that can track protein products without mitigating their function, such as the addition of small sequences which will serve as binding motifs to monoclonal antibodies.

FINAL POLICY STATEMENT FOR RESEARCH AND REGULATION OF BIOTECHNOLOGY PROCESSES AND PRODUCTS

I. Introduction

The USDA portion of the "Proposal for a Coordinated Framework for Regulation of Biotechnology" (hereafter referred to as the December 3u, 1984 Notice) appeared at 49 FR 50897-50904. As a part of its policy perspective, USDA stated that agriculture and forestry products developed by biotechnology will not differ fundamentally from conventional products and that the existing regulatory framework is adequate to regulate biotechnology.

USDA has both research and regulatory responsibilities for biotechnology activities. This document provides significant new information in both areas. Section II describes 1985 Federal Register notices concerning USDA policies and responsibilities for biotechnology. Included in this discussion is an explanation of the assignment of responsibilities within USDA for the oversight of USDA funded research and for the regulation of the products of biotechnology. An understanding of the way in which USDA has divided these responsibilities should prove helpful to those in the private sector seeking review and/or approval of biotechnology applications.

A new section III has been added describing USDA's policy for agricultural biotechnology research. USDA is publishing as a companion document, USDA Guidelines for Biotechnology Research that will closely parallel the NIH Guidelines. The USDA guidelines will be issued under the authority of the Food Security Act of 1985 (Pub. L. 99-198). This Act amended section 1404(2) of the National Agriculture Research, Extension, and Teaching Policy Act (NARETPA). The Amendment gave the Secretary of Agriculture responsibility for establishing "appropriate controls with respect to the development and use of the application of biotechnology to agriculture." All USDA funded agriculture biotechnology research or research conducted at an entity receiving USDA funds would be subject to the USDA Guidelines for Biotechnology Research unless the specific research project is supported by and subject to the guidelines or regulations of another Federal agency. These Guidelines would encompass all phases of agricultural biotechnology research, i.e. (1) Contained laboratory experiments; (2) specialized isolation research (e.g., greenhouse, biotron); and (3) environmental research release (e.g., controlled and segregated field plots). USDA hopes that entities not required to comply with the Guidelines would voluntarily adhere to the requirements. To encourage compliance, USDA proposes to adopt the NIH policy of providing the researchers not required to comply with these Guidelines the opportunity to have their new biotechnology research proposals reviewed by USDA.

Those entities covered by the USDA Guidelines for Biotechnology Research would also be required to comply with any applicable statutes such as those set forth in section IV of this document, and any regulatory issues thereunder.

The Secretary of Agriculture has established an Office of Agriculture Biotechnology (OAB), which will have primary responsibility for implementing and coordinating the Department's policies and procedures pertaining to all facets of biotechnology. This includes the conduct of laboratory and field research, exprimentation on biotechnology products prior to their commercialisation, and all matters of oversight of biotechnology in agriculture. The new office will report to the Assistant Secretary for Science and Education through the authority provided in the amendment to the Food Security Act of 1985. The Assistant Secretary for Science and Education will seek to establish an Agriculture Biotechnology Recombinant DNA Advisory Committee (ABRAC) and shall continue the responsibilities for agriculture formerly handled by the NIH-RAC during the last 10 years. The OAB shall operate in a close parallel manner to the Office of Recombinant DNA Activities (ORDA) of the National Institutes of Health. This includes the responsibility of the ABRAC and the implementation of the USDA Guidelines for Biotechnology Research. The NIH system is well respected both domestically and worldwide, and has achieved a high degree of efficiency in achieving broad confidence in the safety of new biological research conducted under its requirements.

The OAB also will serve as a focal point for coordinating a National Biological Impact Assessment Program, which is to evaluate and monitor the potential impacts of biotechnological processes and products on safety and the environment.

Section IV contains USDA's regulatory policy statements for veterinary biological products, plants and plant products, meat and poultry products, and seeds. USDA stated in the December 31, 1984 Notice that while its existing regulatory framework is adequate, it would constantly reevaluate its regulatory position and should additional regulatory measures become necessary, amend its regulations (49 FR 50904). For veterinary biologicals regulated under the Virus-Serum-Toxin Act (VSTA), USDA has identified three categories which may be derived by recombinant DNA techniques or developed from hybridomas. The categories are based on biological characteristics and safety concerns, and are described fully in section IV(A). The first category consists of inactivated recombinant DNA-derived vaccines, bacterins, bacterin-toxoids, virus subunits, or bacterial subunits, as well as monoclonal products. This category presents no new or unusual safety or environmental concerns. The second category includes those products containing live microorganisms that have been modified by the addition or deletion of one or more genes. Such products will be evaluated under current regulatory policies and procedures to assure that the addition or deletion of specific genetic information does not impart increased virulence, pathogenicity, or survival advantages. The third category included products using live vectors to carry recombinant derived foreign genes for immunizing antigens and/or other immune stimulants. Characteristics of safety and transmission must be established fully before questions and concerns dealing with safety to humans, animals, and release into the environment can be answered and before such products can be considered for licensing. Section IV(A) also includes new information about revised USDA review procedures for the importation of cell cultures and hybridomas. A brief discussion is included about the proposed regulations implementing the provisions of the amendments to the VSTA contained in the Food Security Act of 1985.

For organisms and products derived by the techniques of genetic engineering, USDA is proposing new rules to regulate organisms which are plant pests or which there is reason to believe are plant pests. It is USDA's policy to regulate certain genetically engineered organisms if the donor, vector/vector agent, or recipient organism is a member of a group of organisms that are known to contain plant pests, or if based on experience, USDA determines that a genetically engineered organism or product is a plant pest or if USDA has reason to believe that a genetically engineered organism or product is a plant pest. The proposed regulations are summarized in section IV(B).

The USDA policy for regulating meat and poultry products and seeds derived through biotechnology remains substantially as stated in the December 31, 1984 Notice, and appears in section IV (C) and (D).

A new section (V) has been added describing the scientific review mechanisms to be established by USDA to assit USDA Agencies in biotechnology research and regulatory decision-making. USDA has established a Committee on Biotechnology in Agriculture (CBA) chaired by the Assistant Secretary for Science and Education and the Assistant Secretary for Marketing and Inspection Services.

A detailed summary of comments on the December 31, 1984 Notice and USDA responses appears as section VI. The comments are organised to conform to the form of the December 31, 1984 Notice, with general comments and responses on the USDA regulatory philosophy followed by comments and responses on specific aspects of USDA's regulatory structure.

II. Notices

Three Federal Register notices concerning the Department's biotechnology related activities have been published subsequent to publication of the December 31, 1984 Notice.

On July 19, 1985, a document amending the delegations of authority of USDA to assign responsibility for these Research and regulatory activities (7 CFR Part 2) was published in the Federal Register (50 FR 29367-29368).

In this document, the Secretary of Agriculture delegated responsibility to the Assistant Secretary for Marketing and Inspection Services to coordinate the development and carrying out of all matters and functions pertaining to the Department's regulation of biotechnology and to act as liaison on all matters and functions pertaining to the regulation to biotechnology between agencies within the Department and between the Department and government and private organisations. These responsibilities were further delegated from the Assistant Secretary for Marketing and Inspection Services to the Administrator of the Animal and Plant Health Inspection Service (APHIS).

Also in this document, the Secretary of Agriculture delegated responsibility to the Assistant Secretary for Science and Education to coordinate the development and carrying out of all matters and functions pertaining to agricultural research involving biotechnology conducted or funded by the Department including the development and implementation of guidelines for oversight of research activities, and to act as liaison on all matters and functions pertaining to agricultural research in biotechnology between agencies within the Department and between the Department and other governmental, educational and private organisations. n1

n1 The Assistant Secretary for Science and Education oversees the research activities of the Agricultural Research Service (ARS), the Cooperative State Research Service (CSRS), the Extension Service (ES), and the Office of Grants and Program Systems (OGPS). The Assistant Secretary for Marketing and Inspecting Services oversees the regulatory activities of the Animal and Plant Health Inspection Service (APHIS), which includes Veterinary Services (VS) and Plant Protection and Quarantine

(PPQ); the Agricultural Marketing Service (AMS); and the Food Safety and Inspection Service (FSIS). The policies and procedures of these agencies for biotechnology were described in the USDA portion of the coordinated policy statement at 49 FR 50899-50904 On September 23, 1985, USDA's APHIS published a notice which contained its policy statement and requirements for the control and protection of documents that contain confidential business information concerning biotechnology and the veterinary biologics program (50 FR 38561-38563).

On November 14, 1985, the Office of Science and Technology Policy published a notice in the Federal Register announcing the establishment of the Biotechnology Science Coordinating Committee (BSCC) (50 FR 47174-47195). This Committee is to serve as an interagency forum for coordinating science issues related to research and commercial applications of biotechnology.

The notice also stated that USDA will establish a Committee on Biotechnology in Agriculture (CBA) to assit in assuring that research and regulatory decisions and based on the best science available.

III. USDA Research Policy Statement

USDA supports research to promote and protect the general health and welfare of the people of the United States. n2 Research program include: Studies on production of food and agricultural processing and marketing; identity and development of new crop and animal sources of food, fiber, and energy; increased agricultural efficiency and reduction of dependence on petroleum-based products; development of improved management and conservation of soil, water, forest, and range resources. The programs are fulfilled through State, Federal, and private industry cooperative efforts.

In the areas of agricultural research relevant to biotechnology, many plant, animal, and microbial alterations have been developed for release through traditional genetic approaches such as mutagenesis and hybridisation. In a complementary vein, beneficial introduction of organisms from abroad have established a sound base for research and regulatory oversight.

The experience with these bases provide a substantial knowledge base for conducting evaluations of the safety and efficacy of biotechnology processes and products.

USDA will evaluate the environmental impacts in the context of individual experiments that encompass the entire range of experimentation from contained facilities to open field testing. As knowledge and experience are gained, broadly applicable procedures and guidelines will be developed. Particular consideration will be given to the stability of engineered changes and the possibility that genetic elements might be transferred from one organism to another. Also important will be the development of data that will enable predictions of which organisms may become established in new ecosystems, and resulting environmental consequences.

USDA considers products developed through biotechnological techniques as no different from those products resulting from research using conventional techniques providing appropriate research review is conducted with established protocols. Agricultural biotechnology research activities require appropriate review to avoid untoward effects on human health and the environment.

USDA expects to rely on the existing network of scientific expertise in the agriculture research community. Thousands of plant selections, animal breeding lines, and microorganisms are tested annually at sites under varying climatic conditions through the Nation. This network of scientific expertise permits continual, open assessment of agricultural research and products of that research in the field. USDA has broad statutory authority to conduct and support research in wide ranging areas of agriculture. In addition to the authorities described in the matrix of Federal Laws related to biotechnology found in the Federal Register Notice of November 14, 1985 (50 FR 47174-47195) the Food Security Act of 1985 (Section 1404(2) of the National Agriculture Research, Extension, and Teaching Policy Act Amendments of 1985, Pub. L. No. 99-198), made the Secretary of Agriculture responsible for establishing "appropriate controls with respect to the development and use of the application of biotechnology to agriculture." Through this authority, and pursuant to the Delegation of Authority Pertaining to Biotechnology published in the Federal Register on July 19, 1985 (50 FR 29367-68), the Assistant Secretary for Science and Education will complete development of a national system of agricultural biotechnology research oversight in much the same manner that agriculture has been a part for the last 10 years through the NIH-RAC.

The Assistant Secretary for Science and Education has initiated the establishment of the Agriculture Biotechnology and Recombinant DNA Advisory Committee (ABRAC), to be managed through an Office of Agriculture Biotechnology (OAB) which is a parallel to the National Institutes of Health Recombinant DNA Advisory Committee (NIT-RAC) and Office of Recombinant DNA Activities (ORDA). The OAB will serve as the focal point for developing and coordinating USDA policies and activities pertaining to biotechnology research and will perform related interagency and public liaison functions. OAB will also assist in carrying out the responsibilities assigned to the Assistant Secretary for Science and Education, including the development and implementation of policies and procedures, and guidelines for the conduct of laboratory and field research.

All federally-funded agriculture biotechnology research or research conducted at an entity receiving USDA funds will be subject to the USDA Guidelines for

Biotechnology Research, which are published as a companion document to this policy statement, unless the specific research project is supported by and subject to the guidelines or regulations of another Federal agency. These Guidelines encompass the entire spectrum of degrees of containment in agricultural biotechnology research i.e.: (1) Contained laboratory experiments; (2) specialized isolation research (e.g., greenhouse, biotron); and (3) environmental research agricultural biotechnology release (e.g., controlled and segregated field plots). Research investigators not required to comply with USDA Guidelines will be encouraged to follow these Guidelines. To assure consistency, USDA adopted the model established by the NIH of providing such researchers with the opportunity to have their biotechnology research proposals reviewed as required by the Guidelines.

The USDA Guidelines for Biotechnology Research require that research organisation use the Institutional Biosafety Committee (IBC) concept as established by NIH. This requirement assures that each research organisation and its investigators employ a multidisciplinary team to assit in carrying out their responsibilities under the Guidelines. The IBC's, as described in the Guidelines, would consist of persons with relevant agricultural expertise in areas such as recombinant DNA technology, biological safety, physical containment, and ecology. Requests for review beyond IBC should be sent to the Office of Agriculture Biotechnology (OAB) through the Assistant Secretary of Science and Education, Room 324-A, Administration Bldg., Washington, D.C. 20250.

These Guidelines also would require compliance with existing statutes of the USDA involving the movement of regulated organisms that require the issuance of a permit. The movement of microorganism injurious to plants and animals as well as the movement of certain non-indigenous plants and animals would continue to follow long-established procedures for USDA approval. After review, a permit, if needed, may be issued that allows movement. It is the responsibility of the research scientists to obtain that permit.

The Assistant Secretary for Science and Education will complete establishment of a National Biological Impact Assessment Program (NBIAP) as indicated in the USDA Guidelines for Biotechnology Research. NBIAP would serve to assist USDA in the evaluation and monitoring of biotechnology research and impact over time. Coordination of NBIAP will be provided through OAB.

IV. USDA Regulatory Policy Statements

The existing USDA regulatory authority for biotechnology was listed in the matrix of the December 31, 1984 Notice at 49 FR 50860-50874 and described in brief at 49 FR 50898-50899. The statutes considered most applicable to biotechnology applications are the Virus-Serum-Toxin Act (VSTA) of 1913 (21 U.S.C. 151-158), the

Federal Plant Pest Act (FPPA) of May 23, 1957 (7 U.S.C. 150aa-150jj), the Plant Quarantine Act (PQA) of August 20, 1912 (7 U.S.C. 151-164, 166, 167), the Organic Act of September 21, 1944 (7 U.S.C. 147a), the Federal Noxious Weed Act (FNWA) of 1974 (7 U.S.C. 2801 *et seq.*), the Federal Seed Act (FSA) (7 U.S.C. 551 *et seq.*), the Plant Variety Protection Act (PVPA) (7 U.S.C. 2321 *et seq.*), the Federal Meat Inspection Act (FMIA) (21 U.S.C. 601 *et seq.*), and the Poultry Products Inspection Act (PPIA) (21 U.S.C. 451 *et seq.*).

A. Veterinary Biological Products

Under the Virus-Serum-Toxin Act of 1913, 21 U.S.C. 151-158, the USDA exercises regulatory authority over all veterinary biologics imported into the United States or shipped or delivered for shipment interstate. Recent amendments contained in the Food Security Act of 1985 have extended this authority to products which are shipped intrastate or exported, and have given the Department additional enforcement mechanisms such as the power to detain and seize products. Under the VSTA, veterinary biologics may not be shipped or delivered for shipment if they are worthless, contaminated, dangerous, or harmful. Veterinary biological products must be prepared in a USDA-licensed establishment under regulations promulgated by the Secretary of Agriculture. Those products which are imported into the United States must be imported under a permit issued by the Secretary. The pertinent regulations for veterinary biologics are found in Title 9 of the Code of Federal Regulations, Parts 101 through 117. New regulations will be drafted to implement the provisions of the amendments to the VSTA. Such regulations will provide for a more comprehensive regulatory scheme, including seizure and condemnation and detention procedures. They also will establish procedures to be used in the issuance of special licenses and exemptions provided for by the legislative amendments.

Veterinary biological products are defined in the governing regulations, 9 CFR 101.2(w) as "all viruses, serums, toxins, and analogous products of natural or synthetic origin, such as diagnostics, antitoxins, vaccines, live microorganisms, killed microorganisms, and the antigenic or immunizing components of microorganisms intended for use in the diagnosis, treatment, or prevention of diseases of animals."

Licensing provisions for veterinary biological products and establishments are found in Part 102 of the USDA regulations (9 CFR Part 102). A product license requires the satisfactory completion of various requirements to assure purity, safety, potency, and efficacy of the products. The specific requirements were discussed in the December 31, 1984 Notice at 49 FR 50899.

Pursuant to § 103.3 (a) through (g) of the USDA regulations, a person may be authorized to ship unlicensed biological products for the purpose of evaluating experimental products by treating limited numbers of domestic animals if USDA determines that the conditions under which the experiment is to be conducted are adequate to prevent spread of disease and approves the procedures set forth in the request for such authorisation (9 CFR 103.3 (a)-(g)).

Upon satisfactory completion of all requirements, including review and acceptance of labels, a U.S. Veterinary Biological Product License may be issued.

The application of new biotechnological procedures for the production of veterinary biological products is expanding constantly. For the purposes of licensing, biologics derived by recombinant DNA-techniques or developed from hybridomas, may be classified into three broad categories. This division is based upon the biological characteristics of the new products and the safety concerns they present, and is wholly analogous to the approach used in other veterinary biologics.

The first category includes inactivated recombinant DNA-derived vaccines, bacterins, bacterin-toxoids, virus subunits, or bacterial subunits. These nonviable or killed products pose no risk to the environment and present no new or unusual safety concerns. Monoclonal antibody (hybridoma) products used prophylactically, therapeutically, or as components of diagnostic kits also are included in this category.

The second category includes those products containing live microorganisms that have been modified by the addition or deletion of one or more genes. Deleted genes may code for virulence, oncogenicity, enzyme activity, or other biochemical functions. Added genes may result in the expression of new immunizing antigens or the production of novel biochemical byproducts such as beta-galactosidase. Precautions must be exercised to assure that this addition or deletion of specific genetic information does not impart increased virulence, pathogenicity, or survival advantages in these organisms which are greater than those found in natural or wild-type forms.

Modifications also must not impart undesirable new or increased adherence or invasion factors, colonisation properties, or intrahost survival factors. It is important that genes added or deleted do not compromise the safety characteristics of the organisms. In most cases it is expected that they will be improved, and would therefore not pose any new threat to humans, other animal species, or to the environment.

The genetic information to be added or deleted must consist of well-characterized DNA segments. Required licensing data may include base pair analysis, sequence information, restriction endonuclease sites, as well as phenotypic characterisation of the altered organism. A comparison is also required to be made between the genetically engineered organism and the wild-type form with respect to biochemical pathways, virulence traits, or other factors affecting pathogenicity.

The third category includes products using live vectors to carry recombinant-derived foreign genes that code for immunizing antigens and/or other immune stimulants. Live vectors may carry multiple recombinant-derived foreign genes since they can carry large quantities of new genetic information. They also are efficient at infecting and immunizing target animal species. These properties, for example, make vaccinia virus recombinants very popular subjects for vaccine development programs.

Live vectors currently being evaluated by licensees, applicants, and other research organisations include vaccinia, bovine papilloma virus, adenoviruses, Simian Virus-40, and yeasts. Characteristics of safety and transmission must be examined before questions and concerns dealing with safety to humans, animals, and release into the environment can be answered and before such products can be considered for licensing.

USDA will continue to avail itself of additional expertise from the Public Health Service "Interagency Group to Monitory Vaccine Development, Production, and Usage." This interagency committee will be utilized to consider potential human health hazards from the use of veterinary biological products and to review issues such as those arising from the potential effect of organisms potentially pathogenic to people or animals.

Veterinary biological products prepared using modern biotechnological procedures such as recombinant DNA, chemical synthesis, or hybridoma technology will be treated similarly to products prepared by conventional techniques. The unlimited number and kind of products that may result from these modern biotechnology procedures make it impossible to define all requirements in specific terms. Each product is evaluated individually to determine what will be necessary to establish its purity, safety, potency, and efficacy. Scientific considerations may dictate generic areas of concerns or the use of certain tests for specific situations. Special assays, preferably using in vitro methods, may be required for potency and stability determinations. Additional tests may be required to assure safety, especially when live microorganisms are present in the biological products.

USDA is authorized to issue three types of permits for importing biological products into the United States (9 CFR 104.2). A separate United States Veterinary Biological Product permit is required for each shipment of biological product to be imported.

Permits are required for imported biological products used for research and evaluation, distribution and sale, or transit shipment only. Requests for application (U.S. Form 14-5) should be submitted to the Veterinary Biologics Staff, Veterinary

Services, Animal and Plant Health Inspection Service, 6505 Belcrest Road, Hyattsville, Maryland 20782.

To provide guidance to current or prospective manufacturers employing modern biotechnological methods, the following points are presented:

1. *Recombinant DNA-Derived Products.* Genetic information coding for a product of interest and other sequences not indigenous to the host are referred to as foreign DNA. Recombinant DNA technology encompasses the isolation, characterisation, and expression of foreign DNA in organisms or vectors. The specific cloned nucleotide segment coding for the desired product or other foreign DNA segments must be defined in data supporting each license application. These data must also include a description of the source of the DNA and the nucleotide sequence.

A vector is a cloning vehicle which provides a suitable origin of replication necessary for production of foreign DNA. Such replicons may be derived from plasmids, bacteriophages or viruses such as vaccina, bovine pipillomavirus, adenoviruses, or SV-40.

Production of functional gene products depends on the efficient expression of cloned DNA-vector complexes in suitable host organisms. Tissue culture cells, bacteria, yeasts, and virus cells may be used as hosts for replication of vectors. The mechanisms of transfer, the copy number, and the physical state of the constructed vector inside the host cell, integrated or extrachromosomal, must be described.

USDA's licensing procedure for veterinary biological products derived from recombinant DNA involves a careful evaluation of each product on an individual basis to assure purity, safety, potency, and efficacy. Scientific and safety considerations may require specific safeguards and procedures in some situations. The USDA strongly recommends that all applicants establish Institutional Biosafety Committees which follow applicable provisions of the NIH Guidelines for Research Involving Recombinant DNA Molecules. USDA intends to propose guidelines which specifically relate to veterinary biological products. Amendments of the regulations and standards dealing with veterinary biologics will also be considered.

2. *Chemically Synthesized Antigens.* When the product consists of chemically synthesized polypetides, the appropriate amino acid sequences will mimic the antigenic site or epitope found in the native antigen where one exists. Procedures used to increase or prolong an immune response, such as coupling to carrier proteins or addition of adjuvants, must also be described. Immunological data derived from chemically synthesized peptides must be as definitive as those from natural antigens.

3. *Monoclonal Antibody Products.* The specificity and potency of monoclonal

antibody will be compared with those of similar policyclonal antibody products where appropriate. The sensitivity and specificity of monoclonal antibody products used in diagnostic test kits and their potency characteristics when used therapeutically must be similar to conventional antibody. Monoclonal antibody must be derived from Master Cell Stocks which meet the applicable requirements of 9 CFR 113.52. In addition, as is currently required, a description of cell cloning procedures, preparation, and characterisation of cell passages must also be provided.

The Outline of Production must describe all processes including scale-up, ascites fluid or cell culture supernatant preparation, purification, concentration, and inactivation, Mouse colonies must be screened to demonstrate freedom from adventitious agents, especially those detected by the mouse antibody production (MAP) test. If the MAP test discloses the presence of adventitious agents, the product shall not be released unless inactivation procedures approved by Veterinary Services have been performed and tests conducted to ensure proper application of the procedures.

4. *Master Seeds.* Bacterial or viral seed stocks used to prepare veterinary biological products must meet established procedures used to certify Master Seeds for biological products.

The Master Seed for recombinant DNA-derived products may consist of a plasmid or virus carrying the inserted gene. This constructed plasmid is then introduced into the appropriate eukaryotic or prokaryotic expression system selected for vaccine production. Genomic DNA may also be transfected directly into a variety of mammalian cells. Alternatively, in such cases, the stable transfected cell could be considered as the Master Seed.

The establishment of Master Seeds consisting of constructed plasmids or transfected cells requires submission of background information concerning the recombinant DNA procedures used to isolate, purify, and identify genetic material from one source and the modification used for inserting of this material into a new host. Data from cloning, isolation, proliferation, and selection of genetically unique cells would be retained by licensed applicants. In order to characterize adequately the foreign DNA used to code for a particular antigen, the manufacturer must provide a nucleotide sequence analysis.

Tissue culture-propagated cells from vertebrate animals used for vector propagation and antigen production must meet the requirements of 9 CFR 113.51 or 113.52.

If a Master Seed has been accepted by Veterinary Services for use in a licensed product, further genetic modifications may be approved with reduced requirements for additional host animal efficacy studies.

Each Outline of Production must be prepared in accordance with 9 CFR 114.9. Outlines must include procedures to ensure consistency in production and recovery of specific antigenic material. Recovery procedures must include removal of excessive antibiotic levels (9 CFR 114.10) and undesirable fermentation byproducts such as excessive levels of bacterial endotoxins. Serial release tests for purity, safety, and potency will be required. In addition product characterisation tests may be required to demonstrate consistent gene expression.

Organisms and Vectors

Pursuant to the Act of February 2, 1903, (21 U.S.C. 111), and the VSTA, USDA has authority to issue such regulations and take such measures as may be deemed proper to prevent the introduction or dissemination into the United States for the contagion of any contagious, infectious, or communicable disease of animals and/or live poultry from a foreign country into the United States or from one State or territory of the United States or the District of Columbia to another. The importation into the United States or interstate shipment of organisms and vectors is regulated under 9 CFR Part 122. Organisms and vectors are defined in 9 CFR 122.1 as entities which may introduce or disseminate any cantagious or infectious disease of animals. Such substances may not be shipped interstate or imported without a permit. Permit applications must completely describe the substances, intended use, location of the permittee, and safeguards.

A number of revised administrative and technical provisions have been instituted to expedite the USDA review and issuance of permits for importation or organisms and vectors which include cell cultures and hybridomas. No animal-origin biological materials, such as cell cultres, monoclonal antibodies, organisms, vectors, or related material, may be imported into the United States without a Veterinary Services (VS) Permit (VS Form 16-3A). To obtain a permit, an application (VS Form 16-3) should be submitted to: Import-Export Staff, Organisms and Vectors, VS, APHIS, USDA, 6505 Belcrest Road, Hyattsville, MD 20782. This is different from the permit required to import veterinary biologics pursuant to Part 104 of the USDA regulations governing such products (VS Form 14-5 and 14-6).

Applicants must also complete the questionnaire entitled "Importation Information" and submit it with their application. Based upon the information submitted by the applicant, a determination will be made if the material to be imported requires safety testing to ensure it is free from livestock pathogens. Safety testing is conducted at the Foreign Animal Disease Diagnostic Laboratory (FADDL), Plum Island, New York.

Applicants will be advised if a safety test is required and will be given an estimate of the cost for conducting the test. Applicants desiring to have material safety tested must enter into a Cooperative Trust Fund Agreement with APHIS, VS, and deposit in advance sufficient funds to cover the estimated cost. The Import-Export Animals and Products Staff will initiate the Cooperative Trust Fund Agreement. In order to expedite the procedure, VS may issue a permit for the material to be shipped to FADDL pending receipt of the funds and Cooperative Trust Fund Agreement. However, the signed Cooperative Agreement, plus the necessary funds, must be received by VS before testing can be scheduled at FADDL.

Usually 60 to 90 days is needed for issuing a permit for importing material to Plum Island, New York, the completion of safety tests, and the transfer of the imported material to the applicant. A minimum of four vials, each containing at least 1 million cells from a uniform lot, is required for the safety testing.

When the test is completed and a determination made that the imported material is free from livestock pathogens, the remainder of the imported material is released directly to the importer under conditions specified in the permit.

If an importer wishes to import cell cultures and/or hybridoma cells on a regular basis, the applicant may enter into a continuous Cooperative Trust Fund Agreement with VS and establish an escrow account to ensure that unnecessary delays will not occur due to insufficient funds.

Each safety test utilizing susceptible host animals usually cost approximately $2,000 to $3,000. Sometimes it is possible to reduce the cost by pooling samples in one host animal test. Scientists at FADDL developed in vitro safety tests to detect certain livestock pathogens resulting in substantial cost savings for importers. The current cost of each in vitro test is approximately $500, depending upon the type of animal disease present in the country of origin as well as the intended use of the imported material.

Safety testing may not be required for some cell cultures imported for human diagnostic purposes and research. Examples of material which could enter without safety testing include cultured human bone marrow cells, amniocentesis samples, and cells imported for karyotype analysis. Applications for such cell cultures will be considered individually.

Permit applications are evaluated by a new classification scheme that correlates intended use of imported cell cultures with the level of safety testing conducted at FADDL.

The following classification of cell cultures is based on intended use and generally indicates the level of safety testing required.

Class I Cell cultures to be used for the production of products such as vaccines, hormones, or other biologicals to be used in livestock, poultry, or for commercial distribution.

Requirement: These cell cultures must be safety tested at FADDL using susceptible host animals, approved in vitro test, and/or laboratory animals.

Class II Cell cultures to be used only for in vitro studies and not to be used in animals other than primates.

Requirement: These cultures may not require safety testing. The material may be sent directly to the importer when no safety testing is required. The permit (VS Form 16-3A) will specify restrictions such as *"For in vitro Laboratory Tests: Do not inoculate into Livestock, Birds, or Laboratory Animals."*

Cell cultures imported under permit which do not require a safety test may not be distributed to other laboratories without prior approval from USDA, APHIS, VS. Applications for the distribution of imported material should be submitted to the USDA, APHIS, VS, Import-Export Staff, Organisms and Vectors.

When appropriate, a review is conducted by the Administrator's Parent Committee on Organisms and Vectors. Members of this committee have wide expertise in evaluating safety. Clearance may also require testing in high security facilities at the Veterinary Services, FADDL, Plum Island, New York.

B. Plants and Plant Products

Pursuant to the authority granted by the Federal Plant Pest Act (FPPA) of May 23, 1957, as amended (7 U.S.C. 150 aa through 150 jj), and the Plant Quarantine Act (PQA) of August 20, 1912, as amended (7 U.S.C. 151 through 164, 166, and 167). USDA has regulatory authority over the movement into or within and through the United States of plants, plant products, plant pests, and any product or article which may contain a plant pest at the time of movement. These articles are regulated in order to prevent the introduction, spread, or establishment of plant pests new to or not widely prevalent in the United States. The regulations implementing this statutory authority are found in 7 CFR Parts 300 through 399.

"Plant Pest," as defined by statute, means any living stage of any insects, mites, nematodes, slugs, snails, protozoa, or other intertebrate animals, bacteria, fungi, or parasitic plants or reproductive parts thereof, viruses, or any organisms similar to or allied with any of the foregoing, or any infectious substances, which can directly or indirectly injure or cause disease or damage in any plants or parts thereof, or any processed, manufactured, or other products of plants (7 U.S.C. 150aa(c)).

"Movement," as defined by statute, means to ship, deposit for transmission in the mail, otherwise offer for shipment, offer for entry, import, receive for transportation, carry, or otherwise transport or move, or allow to be moved, by mail or otherwise (7 U.S.C. 150aa(g)).

The current permit system requirements for the movement into or within and through the United States of plants, plant products, plant pests, and other articles regulated by FPPA and PQA were fully described in the December 31, 1984 Notice at 49 FR 50900-01. The procedures for issuing permits for the movement of plant pests were discussed separately from plants, plant products and other articles which may contain plant pests at 49 FR 50901-02. USDA regulates the importation of noxious weeds through a permit system similar to that established for plant pests. The existing regulations in 7 CFR Part 360 which designate plants as noxious weeds and establish procedures for obtaining an import permit were described at 49 FR 50902.

Regulation of the Introduction of Organisms and Products Altered or Produced Through Genetic Engineering Which Are or Which There is Reason to Believe Are Plant Pests

The FPPA and PQA are applicable to the movement of plants, plant products, and other articles and plant pests developed through genetic engineering if such plants, plant products, other articles, or plant pests present a risk of plant pest introduction, spread, or establishment.

Under the authority granted by the FPPA and PQA, USDA is proposing new regulations which would impose restrictions on the introduction of organisms and products altered or produced through genetic engineering which are plant pests or which there is reason to believe are plant pests.

In accordance with the provisions of the FPPA and PQA. USDA must determine the plant pest status of plants, plant products or articles to be moved into or within or through the United States. The evaluation process for determining what safeguards, if any, can be imposed which would allow the movement of the plant pest without risk that the plant pest would be disseminated were described in the December 31, 1984 Notice at 49 FR 50901-02. For genetically engineered material from dissimilar source organisms (inter-generic combinations), the determination may be complex. Information about genetically engineered organisms produced through the use of donor, vector/vector agent and recipient organisms that are from a list of known plant pests is needed in order that such organisms be properly regulated.

During the past year, USDA has received permit applications to move genetically engineered organisms into or through the United States. USDA is confident that organisms altered through genetic engineering will play a major role in increased plant yield and improved plant quality. However, a genetically engineered organism derived from organisms that are plant pests also presents a risk of plant pest introduction. The organisms themselves, the cultures in which they are transported, or their packaging may be contaminated with plant pathogens. Genetic alteration may create a plant pest new to and not widespread in the United States. It is necessary, therefore, to establish appropriate safeguards to prevent the introduction of genetically engineered organisms that pose a threat to agriculture. Other genetically engineered organisms that are not plant pests or where there is no reason to believe such organisms are plant pests would not be regulated.

New data have to be required in order to properly evaluate permit applications for those organisms which are plant pests or which there is reason to believe are plant pests. A determination was made that additional data requirements would be incorporated into proposed regulations for those genetically engineered organisms which are of concern under the provisions of the FPPA and PQA.

USDA is publishing as a companion document in the "proposed rules section" of this issue of the Federal Register its proposed regulations pertaining to organisms and products altered or produced through genetic engineering which are on plant pests or which there is reason to believe are plant pests.

The proposed regulations would establish a new part entitled, "Introduction of Organisms and Products Altered or Produced Through Genetic Engineering Which Are Plant Pests or Which There is Reason to Believe are Plant Pests", in Title 7 of the Code of Regulations (7 CFR), pursuant to the authority of the FFPA, as amended (7 U.S.C. 150aa-150jj) and the PQA, as amended, (7 U.S.C. 151-164, 166, 167). Such proposed regulations would regulate the importation into and movement within and through the United States as well as prevent the release into the environment of certain organisms, or products altered or produced through genetic engineering, which are plant pests or which there is reason to believe are plant pests.

The proposed regulations would restrict the "introduction" of certain organisms and products altered or produced through genetic engineering, referred to as "regulate articles." In this context, "introduction" means to move into the United States, to release into the environment, or to move interstate, or any attempt thereat." "Release into the environment" means "use of a regulated article outside the constraints of physical confinement that are found in a laboratory, contained greenhouse, or fermenter or other contained structure."

USDA's proposed regulations, which are designed to prevent the release into the environment of genetically engineered organisms which are plant pests or which there is reason to believe are plant pests are consistent with the legislative intent of the FPPA. The FPPA was enacted in 1957 and was intended as "gap filing" legislation for the purpose of protecting American agriculture against invasion by plant pests and diseases which are new to or not theretofore known to be widely prevalent or distributed within and throughout the United States. The FPPA also provides USDA with authority to regulate insects or pests that might later be found to be injurious to cultivated crops. The release into the environment of a genetically engineered plant pest is tantamount to the introduction of a plant pest which is new to and not theretofore known to be widely prevalent within and throughout the United States and subject to regulation under the FPPA.

It should be noted that "regulated article" would be defined as any organism or product altered or produced through genetic engineering, if the donor organism, recipient organism, or vector or vector agent belongs to a group of organisms designated by the proposed regulations as having plant pests or any organism or product which USDA determines is a plant pest or which there is reason to believe is a plant pest. Under USDA's proposed definition, certain microorganisms would be excluded if the recipient microorganism is non-pathogenic, is non-infectious, and otherwise not a plant pest, and resulted from the addition of genetic material that is well characterized and contains only non-coding regulatory regions. Restrictions would be required for regulated articles because they are plant pests, or because USDA has reason to believe they are plant pests. The proposed regulations would require that a person obtain a permit prior to the introduction of a regulated article and would list specific conditions required for the introduction of a regulated article. The regulated article could be introduced only if all conditions in the proposed regulations as well as all conditions specified on the permit were met. It is important to note that in considering whether a permit can be issued for the introduction of a genetically engineered organism, USDA will perform the same comprehensive analysis that is used in determining whether a permit can be issued for the movement of a "conventional" plant pest. Such asessment shall include an examination of the factors that were discussed in the December 31, 1984, Notice at 49 FR 50901-02 as part of the evaluation process for determining what safeguards can be imposed which would allow the movement of a plant pest without risk of dissemination. These factors are oriented toward an examination of the ecological and environmental effects of a release of the genetically engineered organism or product into the environment.

The proposed regulations also contain provisions for a certificate of exemption for those organisms or products altered or produced through genetic engineering that are not subject to the proposed regulations. A person seeking to introduce an exempt article could voluntarily request a certificate of exemption to facilitate the introduction of the organism or product.

The proposed regulations provide a list of groups of organisms which are plant pests or contain plant pests. If the donor, vector/vector agent, or recipient of the genetically engineered organism is derived from an organism on the list of organisms containing plant pests, such genetically engineered organism would be deemed a "regulated article".

As defined in the proposed regulations, a plant pest includes microorganisms such as bacteria and viruses, and thus a "regulated article" may be a microorganism unless it meets the provisions for exclusion. It is important to note that in some instances certain microorganisms will be subject to joint regulation by USDA and EPA. USDA has jurisdication over certain microorganisms under the FPPA and PQA if the microorganisms are a plant pest. EPA would have jurisdiction under the Toxic Substances Control Act (TSCA) if the microorganism is deemed to be a "new" microorganism or under the Federal Insecticide, Fungicide, and Rodenticide Act, as amended (FIFRA) if the microorginism is to be used as a pesticide. Because each Agency has a different statutory mandate, certain jurisdictional overlaps cannot be avoided. However, EPA and USDA will work cooperatively and simultaneously in the evaluation of genetically engineered microorganisms that fall under the jurisdication of both Agencies. To expedite the review of these microorganisms each Agency will appoint contact persons to coordinate the review to ensure data requests are not duplicated.

The specifics of which microorganisms will be subject to dual Agency review, or primarily single Agency review, is set forth in the preamble of USDA's proposed regulations being published as a companion document to this policy statement. That document should be consulted for further information.

A key to determining whether a genetically engineered organism will be regulated by USDA is the list of organisms containing plant pests that appears in § 340.2 in proposed Part 340. USDA acknowledges that this is not an exhaustive list, and that it does not attempt to list every pest species. Comments are welcome on the list as well as on other parts of the proposed regulations.

In order to solicit as many comments as possible on the list and all other parts of the proposed regulations, USDA has scheduled public hearings in Washington, DC and Sacramento, California, during the 60-day comment period. The time and place of the public hearings as well as the address to send written comments is specified in the preamble to the proposed regulations.

USDA believes that through the submission of detailed comments and full participation by public and private interests, USDA will be able to promulgate a final regulation that will prevent the introduction and dissemination of genetically engineered organisms which are plant pests or which there is reason to believe are plant pests, yet not impede the development of biotechnology.

C. Meat and Poultry Products

The Food and Safety Inspection Service (FSIS) is responsible for assuring the safety, wholesomeness, and proper labeling of food products prepared from domestic livestock and poultry. The Federal Meat Inspection Act (FMIA) and the Poultry Products Inspection Act (PPIA) require FSIS to inspect cattle, sheep, swine, goats, equines, poultry, and food products prepared from them which are intended for use as human food to assure that they are wholesome, not adulterated, and properly labeled, marked, and packaged. Inspection under these statutes is mandatory. The cost of inspection, except for overtime and holiday inspection work, is required to be borne by the USDA. Food, animals and animal products, other than those required to be inspected under the FMIA and PPIA, may be inspected under a voluntary, reimbursable inspection program established under the Agricultural Marketing Act of 1946.

Within the framework of food safety statutes, FSIS has developed regulations for research on animals that are administered experimental animal drugs, biologics, and pesticides (9 CFR 309.17 and 381.75). These regulations state that no animal used in any research investigation involving an experimental biological product, drug, or chemical shall be eligible for slaughter at an official establishment unless certain conditions are met. These conditions include any of several different ways of demonstrating that the use of such biological product, drug, or chemical will not result in the products of such animals being adulterated.

Products Subject to Review. FSIS anticipates that many food animals which are subject to the new techniques of modern biotechnology will not differ substantially in appearance, behavior, or general health from currently inspected cattle, sheep, swine, goats, equines, and poultry. They would be subject to the same inspection procedures and regulations as tradionally inspected food animals. FSIS is aware that some genetically engineered animals, such as mosaics, chimeras, and some hybrids, may differ substantially from animals that are inspected currently under the FMIA and PPIA. If such animals are ever intended for use as human food and are presented for inspection at an official establishment, a decision would have to be made as to whether such animals were covered under the FMIA or PPIA, and if not, whether the FMIA and PPIA should be amended to require inspection of such animals and their products.

Implementation of Review Authority. FSIS's approach toward the review of food animals resulting from the techniques of modern biotechnology consists, in general, of two phases. The first, an experimental phase, focuses on the experimental aspects of vector administrative, gene transfer and gene expression. Since artificial vectors used in animal gene transfer may be considered as either animals drugs or animal biologics, their administration to food animals would be covered under the current regulations on animals used for research (9 CFR 309.17 and 381.75). The requirement that an animal carcass intended for use as human food not be adulterated may require that certain phenotypic, biochemical, and microbiological parameters not be exceeded before the animal can be slaughtered for human food. Depending on future developments, FSIS may amend the regulations (9 CFR 309.17 and 381.75) to provide further assurance that the products of animals genetically engineered by certain techniques are not adulterated. The second phase would be carried out under existing regulations (9 CFR Parts 301 through 381) and would focus on the commercial development, production, inspection and labeling of food animals and food animal products.

D. Seeds

The Federal Seed Act (FSA) (7 U.S.C. 1551 *et seq.*) defines USDA regulatory authority over the importation and interstate shipment of agricultural and vegetable seeds. It does not apply to the production or intrastate distribution of seeds or to seeds other than agricultural or vegetable seeds ("agricultrual seeds" are grass, forage, and field crop seeds).

The FSA prohibits interstate shipment of seed that contains noxious weed seeds at levels in violation of the laws of the State of destination or in excess of levels allowed by the Secretary of Agriculture. This provision applies primarily to seed adulterated with noxious weed seed. In a few instances, however, States have determined that a particular variety of agricultural or vegetable seed is itself a noxious weed. In these instances, FSA prohibits the interstate shipment of the seed into those States. The FSA also allows the Secretary to prohibit the importation of agricultural and vegetable seed which is adulterated with noxious weed seed or which is unfit for seeding purposes.

The authority granted to the Secretary by the FSA to prohibit the interstate shipment or importation of seeds which are found to be detrimental to the agricutural interests of the United States applies to seeds genetically engineered with the modern biotechnology to the same extent as any other seeds.

V. Scientific Review Mechanisms

The manner in which both regulation and oversight of research in agriculture-related biotechnology evolves and is implemented in the United States will have a direct impact on the competitiveness of U.S. industry in both domestic and world markets. Inconsistent or unnecessary procedures for regulation and research will place the U.S. scientific effort and U.S. producers at a substantial disadvantage. It also is important that safeguards be built into biotechnological research processes, and that releases be based on careful evaluations while further experience is being gained. Therefore, USDA feels that such regulatory and research decisions must be based on the best science available. While the responsibilities within USDA for biotechnology reside with the Assistant Secretary for Science and Education and the Assistant Secretary for Marketing and Inspection Services as the delegates of the Secretary of Agriculture, in carrying out their respective responsibilities based on the best science available, they would be able to take advantage of the expertise and perspectives within the Federal Government through a committee to be called the Committee on Biotechnology in Agriculture (CBA). The CBA, to be chaired by these two Assistant Secretaries, will function both as a policy body in the USDA and a bridge between its research and regulating structures.

Committee on Biotechnology in Agriculture

The objectives of the CBA will include:

(i) To provide advice, when requested, on initiatives, proposals, and policy for agriculture-related regulation and research, and assist in the coordination of these activities;

(ii) To review scientific issues submitted by agencies within the Department;

(iii) To assist in identifying data gaps for basic research in agricultural biotechnology;

(iv) To foster public awareness of the scientific issues in biotechnology;

(v) To provide Departmental support for participation in the FCCSET BSCC. USDA expects that the CBA also will utilize existing cooperative entities (e.g., other Federal agencies, universities, State regulatory officials, the public sector, and industry) to acquire, when necessary, information for addressing those issues submitted to it. Such entities, when requested, can provide technical support for sound regulatory and research decisions regarding the use of biotechnology in agriculture and foresty. These entities offer a vast scientific resource upon which USDA can draw.

VI. Summary of Comments

USDA received the comments of one hundred-two (102) respondents, one-half of whom commented specifically on the USDA policy statement. Although USDA agencies considered all comments on the coordinated policy proposal, this response is confined to comments on the USDA portion of the notice.

The two largest categories of respondents were business and academic, followed closely by associations representing these interests. Comments came in lesser numbers from environmental and public interest groups, individuals, law firms, and foreign governments, as well as the National Institutes of Health Recombinant DNA Advisory Committee (NIH-RAC) and a member of the U.S. Congress

The USDA response to the comments follows the form of the original notice, with a discussion of comments on regulatory philosophy followed by a response to comments on the regulatory framework.

Comments on the Nature of Products of Modern Biotechnology: Fourteen respondents stressed their agreement with the USDA statement that "agriculture and forestry products developed by modern biotechnology will not differ fundamentally from conventional products," while six commenters dissented. Three respondents felt that genetic engineering across species barriers did create a potentially different product and the possiblity of unique ecological effects. Concern about the "need for public trust" and public assurance on safety and ethical issues was stressed by three commenters. Seven respondents agreed with USDA that "to date, no unique or safety problems have been associated with products of genetic engineering," but four of the same commenters who view biotechnology products as fundamentally different from conventional products stressed that the potential exists for safety problems with biotechnology applications.

Response: USDA recognizes the importance of ecological effects and the need for developing procedures responsive to public concerns about safety.

Although USDA's regulatory philosophy remains as stated, additions to regulatory procedures are being proposed for genetically engineered plants and plant products and veterinary biologics produced by biotechnology (see section IV). The previously discussed delegations of authority within USDA for biotechnology increase the effectiveness of the administration of current and proposed regulatory procedures affecting the products of modern biotechnology. For veterinary biological products, USDA is currently developing additional procedures pursuant to the VSTA, as amended, for evaluating requests to conduct experimental field trials with live vectors containing genetically engineered organisms or to support product license applications. The procedures being developed consider the parental organism and the effect of the gene alteration on the genetic properties of the recipient, especially the survival, reproduction, and dispersal characteristics. A careful analysis of the genetics, biology, and ecology of the wild-type and modified microorganisms will provide as reasonable prediction of the risks which might be associated with use of the altered organisms.

USDA is proposing regulations pursuant to the Federal Plant Pest Act (FPPA) and the Plant Quarantine Act (PQA) for regulating the introduction of certain organisms of products thereof altered or produced through biotechnology which are plant pests or may become plant pests. This proposed rule should assist USDA in assessing the ecological effects of the release of such genetically engineered organisms into the environment.

Guidelines for oversight of agricultural biotechnology research funded by USDA will be issued under the authority of the Food Security Act of 1985.

USDA also is establishing scientific review mechanisms to assist in research and regulatory decisions (see section V).

These proposed modifications in the procedural framework are described as a part of the final policy statement for veterinary biologics, plants and plant products, research, and scientific review mechanisms.

Comments of the Adequacy of Existing Authority: Thirteen commenters agreed with USDA that its existing regulatory framework is adequate for biotechnology applications, and nine favor the case-by-case approach under existing authority. Five commenters felt that new legislation is or may be needed; two of the five oppose the case-by-case approach.

Response: USDA has examined its statutory authority for regulating biotechnology products and processes, and USDA agencies have processed licensing and permit applications under the existing statutes. The existing authority is considered adequate at this time. Established procedures, with the proposed modifications, can be adapted effectively to handle biotechnology applications. USDA is currently considering genetic engineering applications on a case-by-case basis using existing authority.

Comments on Need for Procedures and Guidelines: Sixteen respondents commented that USDA had not outlined procedures for the review and approval of genetically engineered products. Twelve respondents stressed the need for flexibility, and six requested sunset provisions in USDA biotechnology regulations.

Response: The USDA policy statement of December 31, 1984, did outline procedures currently used for the review and approval of certain genetically engineered products. In considering license applications for genetically engineered veterinary biologics, USDA follows the standards and procedures applicable to all such products found in § § 101-117 of the applicable regulations and standards (9 CFR 101-117). In the December 31, 1984 Notice, USDA offered supplementary guidelines for licensing such products. New procedures are being developed to evaluate production and testing of veterinary biologics derived through use of genetic engineering techniques. The information needed for proper evaluation will depend on the parent organism and the effect of the gene alteration on the genetic properties of the recipient. A paper describing the USDA licensing policy for biologics produced by recombinant

DNA technology was presented at the Joint International Association of Biological Standardisation/World Health Organisation Symposium on "Standardisation and Control of Biologics Produced by Recombinant DNA Technology," Geneva, Switzerland, 1983 (published in *Developments in Biological Standardisation,* V. 59, pp. 167-173, S. Korgel, Basel, 1985). The paper describes requirements for plasmid/vector characterisation and stability, and correlation to conventional Master Seed concepts, as well as methodology which can be used to monitor antigenic expression, concentration, purification, and stability testing during production and recovery.

The movement of genetically engineered products which are plant pests and present a risk of plant pest introduction or spread is regulated by 7 CFR 330.200 implemented pursuant to the FPPA and PQA. The movement of organisms and vectors which may cause desease in animals is regulated under 9 CFR Part 122.

USDA realised that the statement left unanswered some questions about the means for review and approval of various genetically engineered products. The proposed regulations described in section IV(B), implemented under the authority of the FPPA establish permit requirements for the "introduction" of organisms altered or produced by genetic engineering which are or may become plant pests. The regulations would be flexible because organisms determined not to be plant pests would be exempt, and this category could be expanded in the future to include organisms whose plant pest status is currently uncertain and therefore restricted. It is hoped that the discussion in section IV(B) of this policy statement answers any remaining questions about the review and approval procedures for such genetically engineered products.

Comments on Confidential Business Information (CBI): Six commenters representing business and scientific interests expressed concern about the protection of "confidential business information" in the USDA regulatory process while two public interest groups stressed the "public's right to know."

Response: The USDA regulations implementing the Freedom of Information Act (FOIA) (5 U.S.C. 552) are found in 7 CFR 1.1-1.16. The FOIA provides that Federal agencies must make available to the public all records not specifically exempt from disclosure. Exemptions include "trade secrets and commercial or financial information." (5 U.S.C. 552(b)(4)). On September 23, 1985, USDA's APHIS issued a policy statement on the protection of privileged or confidential information (50 FR 38561-38563). This policy statement establishes requirements for the control and protection of documents received by APHIS that contain privileged or confidential business information concerning biotechnology and the veterinary biologics program. The procedures established conform to the FOIA requirements for both protection and disclosure.

Comments on Use of NIH Guidelines: Four respondents questioned the USDA requirements that manufacturers of veterinary biological products using recombinant DNA technology follow the National Institutes for Health Guidelines for Research Involving Recombinant DNA Molecules (NIH Guidelines). One respondent thought USDA implied that all people "who work with recombinant DNA plants" would be required to comply with the NIH Guidelines, and requested procedural guidelines for industry.

Response: The USDA does not require that the manufacturers of veterinary biological products or plant products of recombinant DNA technology follow the NIH Guidelines. However, USDA strongly recommends that all license applicants for veterinary biologics follow appropriate provisions of the NIH Guidelines, such as those regarding the establishment of an institutional biosafety committee. USDA intends to propose guidelines that will parallel closely the NIH Guidelines, and it intends to recommend strongly that entities not required to follow the USDA guidelines do so voluntarily.

Comments on Importation of Cell-Lines: Three associations representing biotechnology companies requested that USDA take steps to reduce delays in the clearance and testing procedures required for the importation of biotechnology-derived products and cell-lines. On February 12, 1985, the Association of Biotechnology Companies (ABC) delivered a report on USDA importation guarantine issues to the APHIS Parent Committee for Foreign Pathogens and Vectors. This report was an attachment to the ABC comment letter.

Response: The USDA has instituted a number of revised administrative and technical provisions to expedite the issuance of permits for importation of organisms and vectors which include cell cultures and hybridomas. A supplementary questionnaire, designed to insure adequate information on cell cultures and products from recombinant DNA or hybridoma technologies, now accompanies each permit application. Applicants are advised whether or not a safety test is required and a cost estimate is given. Safety testing may be conducted concurrently with the administrative review of the permit application, but only at APHIS' Foreign Animal Disease Diagnostic Laboratory (FADDL) at Plum Island, New York. New test procedures have reduced the cost of safety testing, and the cost per sample can be further reduced by conducting a safety test with pooled samples. Permit applications are evaluated using a new classification scheme that equates intended use of imported cell cultures with the level of safety testing required at FADDL. Class I cell cultures, employed in the preparation of products such as enzymes, vaccines, or hormones for commercial use, are subject to complete safety testing. Class II cell cultures, used only for in vitro studies and not to be used in animals other than primates, are subject to a lesser degree of testing.

Comments on Risk Analysis: Seven respondents discussed the issue of risk assessment or risk/benefit analysis of biotechnology applications. Comments varied from a recommendation that "standard risk assessment methodologies" be adopted by all agencies to a warning against attempting to regulate the "hypothetical and imaginary "potential' dangers" of recombinant DNA techniques.

Response: The National Environmental Policy Act (NEPA) applies to USDA actions. The "APHIS Guidelines Concerning Implementation of NEPA Procedures" (44 FR 50381, August 28, 1979) would be used to make an environmental assessment or environmental impact statement of the effects of a proposed release of a genetically engineered organism regulated pursuant to the VSTA, the FPPA and PQA, and related statutes. A formal risk management procedure based on a wide variety of safety concepts will be used to evaluate systematically proposed releases. The information required by any new regulations promulgated under the FPPA and PQA would be used to prepare the environmental assessment for release of a genetically engineered product which is a plant pest or may become a plant pest.

In normal husbandry and laboratory practices, veterinary biological products normally are not considered to be released into the environment. In the event that a conventionally prepared or recombinant derived product would be considered to be released into the environment, the issuance of a license or import permit would require compliance with procedures being developed and interagency approval. The procedures under development consider the parental organism and the effect on the gene alteration on the genetic properties of the recipient, especially the survival, reproduction and dispersal characteristics.

Safety, ethics, and policy issues in agricultural biotechnology research will be overseen by the Committee on Biotechnology in Agriculture (CBA) and such supporting technical advisory groups as may be established by the USDA agencies. Currently, all USDA and USDA-sponsored research involving recombinant DNA must be cleared prior to initiation for compliance with the NIH Guidelines.

Comments on Jurisdiction: The potential for overlapping jurisdiction in the policy notice drew the largest number of comments. Eighteen respondents pointed out that both USDA and EPA propose to regulate agricultural microorganisms. Respondents representing the interests of the veterinary biologics industry contended that a jurisdictional dispute between USDA and FDA delayed the approval of bovine interferon. While generally supporting the concept of the memorandum of understanding (MOU) between USDA and FDA to resolve jurisdictional disputes, one respondent challenged the legality of the MOU, noting that it contains the statement that "animal biological products generally act through a specific immune process," while USDA's current regulations do not restrict its jurisdiction to products operating through such a mechanism of action. Industry respondents also pointed out that the intrastate producer of veterinary biologics is not regulated by USDA. Two firms and one industry association urged prompt Federal oversight action so that States do not act independently to regulate biotechnology products.

Response: USDA agrees that there is the potential for overlapping jurisdiction among the Federal agencies involved in regulating biotechnology products. USDA and EPA representatives have discussed jurisdiction over genetic engineering applications since 1983. USDA and EPA have begun to establish a regulatory procedure for reviewing certain submissions of genetically engineered microorganism applications, a procedure which has resulted in joint consultation on several proposals for release into the environment of organisms altered by genetic engineering.

For veterinary biologics regulated under the VSTA, use of procedures currently under development will increase USDA effectiveness in evaluating biotechnology license and product applications. The MOU between USDA and FDA was published on June 8, 1982, in an attempt to resolve the issue of new products which fall into the questionable definitional area between animal drugs regulated by FDA and animal biologics regulated by USDA. An interpretation by some that the term animal biologics only includes substances that act through a specific immune process has resulted in some confusion. There is nothing in USDA's current regulations or law which restricts its jurisdiction to products acting solely through this mechanism of action, and because of this fact, the memorandum qualifies its reference to specific immune process by the word "generally." Although efforts will be made to clarify the issue further, it should be noted that there appears to be little uncertainty about whether a particular product is a veterinary drug or biologic.

The Food Security Act of 1985 contains amendments to the VSTA that extend USDA's jurisdiction to veterinary biologics which are shipped intrastate or exported. The provisions of the amendments are discussed more fully in Section IV.

Comments on the National Biological Impact Assessment Program (NBIAP): Seven respondents commented on the NBIAP, the proposal by the National Association of State Universities and Land Grant Colleges (NASULGC) for establishing a program to assess genetically engineered organisms before they are released into the environment. Three commenters — a member of Congress, a spokesperson for a biotechnology firm, and an officer of an environmental organisation — posed questions about the proposal. The questions concerned the NBIAP's statutory or regulatory status; its relation to other USDA agency operations and other Federal agency operations; the processes of risk assessment to be used; its adequacy to review an increasing volume of products; and the appropriateness of biohazard committees as vehicles for review of commercial processes and products. Four respondents representing NASULGC institutions endorsed the proposal stating the view that the agricultural research community has the capability to develop guidelines and assess impacts of biotechnology research and commercial products. The major goal of the program was thought to be insuring the safety of society and the environment.

Response: NBIAP is a scientific advisory system that would be available to the Assistant Secretary for Science and Education. By this system the USDA can draw upon the best experience available from scientists in universities, Federal laboratories, and industry to help assess the risks involved in the processes and products from RDNA work in biotechnology.

NBIAP shall act in an advisory capacity and is in no direct way a part of the formal approval process. It is available to provide assessment, but is not a mandatory process.

Comments on Definitions, Terms, and Data Requirements: Five respondents recommended changes in the definitions, terms, data requirements or classification used by USDA in the notice. Each recommendation is discussed below.

Two respondents commented on the USDA statement of licensing policy for veterinary biologics produced by modern biotechnical methods at 49 FR 50899-50900. Under the heading "1. Recombinant DNA-Derived Products," a manufacturer of veterinary biologics questioned the need to provide the entire nucleotide sequence of a foreign DNA being cloned into a vector.

It is USDA's position that in order to characterize adequately the foreign DNA used to code a particular antigen, the manufacturer should provide a nucleotide sequence analysis. The construction of the vector used for expression of the cloned nucleotide sequence also should include source and function of the component parts of the vector, i.e., origin of replication, antibiotic resistance genes, promotor, enhancers, etc. The manufacturer also questioned the data requirement under the heading "2. Chemically Synthesized Antigens" concerning the persistence of the immune response following administration of the synthetic peptide. The USDA feels that a major concern with the use of synthetic peptides is the development persistence of the immune response. USDA does not intend to require more stringent efficacy data than that necessary to support a veterinary biologic license application employing natural antigens. However, immunological data derived from chemically synthesized peptides must be as definitive as the serological response from natural or nonsynthetic antigens. With respect to the next sentence in the policy statement, an individual respondent proposed a change from the term "antibody response" to "immune response." It is true that the term used in the sentence "Procedures used to increase or prolong an antibody response . . ." is somewhat limiting and can create confusion between B-cell and T-cell response. Therefore, the recommendation to replace "antibody response" with the term "immune response" with the term "immune response" is accepted, since both T-cell responses as well as T-cell/B-cell interactions would be included in the statement.

On the subject of plants and plant pests, a plant pathologist commented on the references to *Pseudomonas syringae* as plant pathogens under the heading "ice nucleation negative bacteria" at 49 FR 50902. The respondent noted that none of the strains of *Pseudomonas syringae* currently proposed for use are plant pathogens and that it would be more correct to call *P. syringae* plant-associated bacteria, some of which are pathogens. USDA will clarify future references to these organisms as the respondent suggests. According to current practice, and under the proposed FPPA regulations, an applicant for a USDA permit to import or move *Pseudomonas syringae* would be required to submit data to show whether of not the strain was a plant pest.

3

Introduction to Gene Science and Science of Genetics

GENE

A gene is a locatable region of genomic sequence, corresponding to a unit of inheritance, which is associated with regulatory regions, transcribed regions and/or other functional sequence regions. The physical development and phenotype of organisms can be thought of as a product of genes interacting with each other and with the environment. A concise definition of gene taking into account complex patterns of regulation and transcription, genic conservation and non-coding RNA genes, has been proposed by Gerstein et al. "A gene is a union of genomic sequences encoding a coherent set of potentially overlapping functional products".

In cells, genes consist of a long strand of DNA that contains a promoter, which controls the activity of a gene, and coding and non-coding sequence. Coding sequence determines what the gene produces, while non-coding sequence can regulate the conditions of gene expression. When a gene is active, the coding and non-coding sequence is copied in a process called transcription, producing an RNA copy of the gene's information. This RNA can then direct the synthesis of proteins via the genetic code. However, RNAs can also be used directly, for example as part of the ribosome. These molecules resulting from gene expression, whether RNA or protein, are known as gene products.

Genes often contain regions that do not encode products, but regulate gene expression. The genes of eukaryotic organisms can contain regions called introns that are removed from the messenger RNA in a process called splicing. The regions encoding gene products are called exons. In eukaryotes, a single gene can encode multiple proteins, which are produced through the creation of different arrangements of exons through *alternative splicing*. In prokaryotes (bacteria and archaea), introns are less common and genes often contain a single uninterrupted stretch of DNA, called a *cistron*, that codes for a product. Prokaryotic genes are often arranged in groups called operons with promoter and operator sequences that regulate transcription of a single long RNA. This RNA contains multiple coding sequences. Each coding sequence is preceded by a Shine-Dalgarno sequence that ribosomes recognize.

The total set of genes in an organism is known as its genome. An organism's genome size is generally lower in prokaryotes, both in number of base pairs and number of genes, than even single-celled eukaryotes. However, there is no clear relationship between genome sizes and complexity in eukaryotic organisms. One of the largest known genomes belongs to the single-celled amoeba *Amoeba dubia*, with over 670 billion base pairs, some 200 times larger than the human genome. The estimated number of genes in the human genome has been repeatedly revised downward since the completion of the Human Genome Project; current estimates place the human genome at just under 3 billion base pairs and about 20,000–25,000 genes.. A recent *Science* article gives a final number of 20,488, with perhaps 100 more yet to be discovered. The gene density of a genome is a measure of the number of genes per million base pairs (called a megabase, Mb); prokaryotic genomes have much higher gene densities than eukaryotes. The gene density of the human genome is roughly 12–15 genes/Mb.

History

The existence of genes was first suggested by Gregor Mendel (1822-1884), who, in the 1860s, studied inheritance in pea plants and hypothesized a factor that conveys traits from parent to offspring. He spent over 10 years of his life on one experiment. Although he did not use the term *gene*, he explained his results in terms of inherited characteristics. Mendel was also the first to hypothesize independent assortment, the distinction between dominant and recessive traits, the distinction between a heterozygote and homozygote, and the difference between what would later be described as genotype and phenotype. Mendel's concept was given a name by Hugo de Vries in 1889, who, at that time probably unaware of Mendel's work, in his book *Intracellular Pangenesis* coined the term "pangen" for "the smallest particle [representing] one hereditary characteristic". Wilhelm Johannsen abbreviated this term to "gene" ("gen" in Danish and German) two decades later.

In the early 1900s, Mendel's work received renewed attention from scientists. In 1910, Thomas Hunt Morgan showed that genes reside on specific chromosomes. He later showed that genes occupy specific locations on the chromosome. With this knowledge, Morgan and his students began the first chromosomal map of the fruit fly *Drosophila*. In 1928, Frederick Griffith showed that genes could be transferred.

In what is now known as Griffith's experiment, injections into a mouse of a deadly strain of bacteria that had been heat-killed transferred genetic information to a safe strain of the same bacteria, killing the mouse.

In 1941, George Wells Beadle and Edward Lawrie Tatum showed that mutations in genes caused errors in certain steps in metabolic pathways. This showed that specific genes code for specific proteins, leading to the "one gene, one enzyme" hypothesis. Oswald Avery, Collin Macleod, and Maclyn McCarty showed in 1944 that DNA holds the gene's information. In 1953, James D. Watson and Francis Crick demonstrated the molecular structure of DNA. Together, these discoveries established the central dogma of molecular biology, which states that proteins are translated from RNA which is transcribed from DNA. This dogma has since been shown to have exceptions, such as reverse transcription in retroviruses.

In 1972, Walter Fiers and his team at the Laboratory of Molecular Biology of the University of Ghent (Ghent, Belgium) were the first to determine the sequence of a gene: the gene for Bacteriophage MS2 coat protein. Richard J. Roberts and Phillip Sharp discovered in 1977 that genes can be split into segments. This leads to the idea that one gene can make several proteins. Recently (as of 2003-2006), biological results let the notion of gene appear more slippery. In particular, genes do not seem to sit side by side on DNA like discrete beads. Instead, regions of the DNA producing distinct proteins may overlap, so that the idea emerges that "genes are one long continuum".

Mendelian Inheritance and Classical Genetics

Darwin used the term Gemmule to describe a microscopic unit of inheritance, and what would later become known as Chromosomes had been observed separating out during cell division by Wilhelm Hofmeister as early as 1848. The idea that chromosomes were the carriers of inheritance was expressed in 1883 by Wilhelm Roux. The modern conception of the gene originated with work by Gregor Mendel, a 19th century Augustinian monk who systematically studied heredity in pea plants. Mendel's work was the first to illustrate particulate inheritance, or the theory that inherited traits are passed from one generation to the next in discrete units that interact in well-defined ways. Danish botanist Wilhelm Johannsen coined the word "gene" in 1909 to describe these fundamental physical and functional units of heredity, while the related word genetics was first used by William Bateson in 1905. The word was derived from Hugo De Vries' 1889 term *pangen* for the same concept, itself a derivative of the word *pangenesis* coined by Darwin (1868). The word pangenesis is made from the Greek words *pan* (a prefix meaning "whole", "encompassing") and *genesis* ("birth") or *genos* ("origin").

According to the theory of Mendelian inheritance, variations in phenotype - the observable physical and behavioral characteristics of an organism - are due to variations in genotype, or the organism's particular set of genes, each of which specifies a particular trait. Different genes for the same trait, which give rise to different phenotypes, are known as alleles. Organisms such as the pea plants Mendel worked on, along with many plants and animals, have two alleles for each trait, one inherited from each parent. Alleles may be dominant or recessive; dominant alleles give rise to their corresponding phenotypes when paired with any other allele for the same trait, while recessive alleles give rise to their corresponding phenotype only when paired with another copy of the same allele. For example, if the allele specifying tall stems in pea plants is dominant over the allele specifying short stems, then pea plants that inherit one tall allele from one parent and one short allele from the other parent will also have tall stems. Mendel's work found that alleles assort independently in the production of gametes, or germ cells, ensuring variation in the next generation.

Prior to Mendel's work, the dominant theory of heredity was one of blending inheritance, which proposes that the traits of the parents blend or mix in a smooth, continuous gradient in the offspring. Although Mendel's work was largely unrecognized after its first publication in 1866, it was rediscovered in 1900 by three European scientists, Hugo de Vries, Carl Correns, and Erich von Tschermak, who had reached similar conclusions from their own research. However, these scientists were not yet aware of the identity of the 'discrete units' on which genetic material resides.

A series of subsequent discoveries led to the realization decades later that chromosomes within cells are the carriers of genetic material, and that they are made of DNA (deoxyribonucleic acid), a polymeric molecule found in all cells on which the 'discrete units' of Mendelian inheritance are encoded. The modern study of genetics at the level of DNA is known as molecular genetics and the synthesis of molecular genetics with traditional Darwinian evolution is known as the modern evolutionary synthesis.

Physical Definitions

The vast majority of living organisms encode their genes in long strands of DNA. DNA consists of a chain made from four types of nucleotide subunits: adenosine, cytidine, guanosine, and thymidine. Each nucleotide subunit consists of three components: a phosphate group, a deoxyribose sugar ring, and a nucleobase. Thus, nucleotides in DNA or RNA are typically called 'bases'; consequently they are commonly referred to simply by their purine or pyrimidine original base components adenine, cytosine, guanine, thymine. Adenine and guanine are purines and cytosine and thymine are pyrimidines. The most common form of DNA in a cell is in a double helix structure, in which two individual DNA strands twist around each other in a right-handed spiral. In this structure, the base pairing rules specify that guanine pairs with cytosine and adenine pairs with thymine (each pair contains one purine and one pyrimidine). The base pairing between guanine and cytosine forms three hydrogen bonds, while the base pairing between adenine and thymine forms two hydrogen bonds. The two strands in a double helix must therefore be *complementary*, that is, their bases must align such that the adenines of one strand are paired with the thymines of the other strand, and so on.

Due to the chemical composition of the pentose residues of the bases, DNA strands have directionality. One end of a DNA polymer contains an exposed hydroxyl group on the deoxyribose, this is known as the 3' end of the molecule. The other end contains an exposed phosphate group, this is the 5' end. The directionality of DNA is vitally important to many cellular processes, since double helices are necessarily directional (a strand running 5'-3' pairs with a complementary strand running 3'-5') and processes such as DNA replication occur in only one direction. All nucleic acid synthesis in a cell occurs in the 5'-3' direction, because new monomers are added via a dehydration reaction that uses the exposed 3' hydroxyl as a nucleophile.

The expression of genes encoded in DNA begins by transcribing the gene into RNA, a second type of nucleic acid that is very similar to DNA, but whose monomers contain the sugar ribose rather than deoxyribose. RNA also contains the base uracil in place of thymine. RNA molecules are less stable than DNA and are typically single-stranded. Genes that encode proteins are composed of a series of three-nucleotide sequences called codons, which serve as the "words" in the genetic "language". The genetic code specifies the correspondence during protein translation between codons and amino acids. The genetic code is nearly the same for all known organisms.

RNA Genes

In some cases, RNA is an intermediate product in the process of manufacturing proteins from genes. However, for other gene sequences, the RNA molecules are the actual functional products. For example, RNAs known as ribozymes are capable of enzymatic function, and miRNAs have a regulatory role. The DNA sequences from which such RNAs are transcribed are known as non-coding DNA, or RNA genes.

Some viruses store their entire genomes in the form of RNA, and contain no DNA at all. Because they use RNA to store genes, their cellular hosts may synthesize their proteins as soon as they are infected and without the delay in waiting for transcription. On the other hand, RNA retroviruses, such as HIV, require the reverse transcription of their genome from RNA into DNA before their proteins can be synthesized. In 2006, French researchers came across a puzzling example of RNA-mediated inheritance in mouse. Mice with a loss-of-function mutation in the gene Kit have white tails. Offspring of these mutants can have white tails despite having only normal Kit genes. The research team traced this effect back to mutated Kit RNA. While RNA is common as genetic storage material in viruses, in mammals in particular RNA inheritance has been observed very rarely.

Functional Structure of a Gene

All genes have regulatory regions in addition to regions that explicitly code for a protein or RNA product. A universal regulatory region shared by all genes is known as the promoter, which provides a position that is recognized by the transcription machinery when a gene is about to be transcribed and expressed. Although promoter regions have a consensus sequence that is the most common sequence at this position, some genes have "strong" promoters that bind the transcription machinery well, and others have "weak" promoters that bind poorly. These weak promoters usually permit a lower rate of transcription than the strong promoters, because the transcription machinery binds to them and initiates transcription less frequently. Other possible regulatory regions include enhancers, which can compensate for a weak promoter. Most regulatory regions are "upstream" — that is, before or toward the 5' end of the transcription initiation site. Eukaryotic promoter regions are much more complex and difficult to identify than prokaryotic promoters.

Many prokaryotic genes are organized into operons, or groups of genes whose products have related functions and which are transcribed as a unit. By contrast, eukaryotic genes are transcribed only one at a time, but may include long stretches of DNA called introns which are transcribed but never translated into protein (they are spliced out before translation). Splicing can also occur in prokaryotic genes, but is less common than in eukaryotes.

Chromosomes

The total complement of genes in an organism or cell is known as its genome, which may be stored on one or more chromosomes; the region of the chromosome at which a particular gene is located is called its locus. A chromosome consists of a single, very long DNA helix on which thousands of genes are encoded. Prokaryotes - bacteria and archaea - typically store their genomes on a single large, circular chromosome, sometimes supplemented by additional small circles of DNA called plasmids, which usually encode only a few genes and are easily transferable between individuals. For example, the genes for antibiotic resistance are usually encoded on bacterial plasmids and can be passed between individual cells, even those of different species, via horizontal gene transfer. Although some simple eukaryotes also possess plasmids with small numbers of genes, the majority of eukaryotic genes are stored on multiple linear chromosomes, which are packed within the nucleus in complex with storage proteins called histones. The manner in which DNA is stored on the histone, as well as chemical modifications of the histone itself, are regulatory mechanisms governing whether a particular region of DNA is accessible for gene expression. The ends of eukaryotic chromosomes are capped by long stretches of repetitive sequences called telomeres, which do not code for any gene product but are present to prevent degradation of coding and regulatory regions during DNA replication. The length of the telomeres tends to decrease each time the genome is replicated in preparation for cell division; the loss of telomeres has been proposed as an explanation for cellular senescence, or the loss of the ability to divide, and by extension for the aging process in organisms.

While the chromosomes of prokaryotes are relatively gene-dense, those of eukaryotes often contain so-called "junk DNA", or regions of DNA that serve no obvious function. Simple single-celled eukaryotes have relatively small amounts of such DNA, while the genomes of complex multicellular organisms, including humans, contain an absolute majority of DNA without an identified function. However it now appears that, although protein-coding DNA makes up barely 2% of the human genome, about 80% of the bases in the genome may be being expressed, so the term "junk DNA" may be a misnomer.

Gene Expression

In all organisms, there are two major steps separating a protein-coding gene from its protein: first, the DNA on which the gene resides must be *transcribed* from DNA to messenger RNA (mRNA), and second, it must be *translated* from mRNA to protein. RNA-coding genes must still go through the first step, but are not translated into protein. The process of producing a biologically functional molecule of either RNA or protein is called gene expression, and the resulting molecule itself is called a gene product.

Genetic Code

The genetic code is the set of rules by which a gene is translated into a functional protein. Each gene consists of a specific sequence of nucleotides encoded in a DNA (or sometimes RNA) strand; a correspondence between nucleotides, the basic building blocks of genetic material, and amino acids, the basic building blocks of proteins, must be established for genes to be successfully translated into functional proteins. Sets of three nucleotides, known as codons, each correspond to a specific amino acid or to a signal; three codons are known as "stop codons" and, instead of specifying a new amino acid, alert the translation machinery that the end of the gene has been reached. There are 64 possible codons (four possible nucleotides at each of three positions, hence 43 possible codons) and only 20 standard amino acids; hence the code is redundant and multiple codons can specify the same amino acid. The correspondence between codons and amino acids is nearly universal among all known living organisms.

Transcription

The process of genetic transcription produces a single-stranded RNA molecule known as messenger RNA, whose nucleotide sequence is complementary to the DNA from which it was transcribed. The DNA strand whose sequence matches that of the RNA is known as the coding strand and the strand from which the RNA was synthesized is the template strand. Transcription is performed by an enzyme called an RNA polymerase, which reads the template strand in the 3' to 5' direction and synthesizes the RNA from 5' to 3'. To initiate transcription,the polymerase first recognizes and binds a promoter region of the gene. Thus a major mechanism of gene regulation is the blocking or sequestering of the promoter region, either by tight binding by repressor molecules that physically block the polymerase, or by organizing the DNA so that the promoter region is not accessible.

In prokaryotes, transcription occurs in the cytoplasm; for very long transcripts, translation may begin at the 5' end of the RNA while the 3' end is still being transcribed. In eukaryotes, transcription necessarily occurs in the nucleus, where the cell's DNA is sequestered; the RNA molecule produced by the polymerase is known as the primary transcript and must undergo post-transcriptional modifications before being exported to the cytoplasm for translation. The splicing of introns present within the transcribed region is a modification unique to eukaryotes; alternative splicing mechanisms can result in mature transcripts from the same gene having different sequences and thus coding for different proteins. This is a major form of regulation in eukaryotic cells.

Translation

Translation is the process by which a mature mRNA molecule is used as a template for synthesizing a new protein. Translation is carried out by ribosomes, large complexes of RNA and protein responsible for carrying out the chemical reactions to add new amino acids to a growing polypeptide chain by the formation of peptide bonds. The genetic code is read three nucleotides at a time, in units called codons, via interactions with specialized RNA molecules called transfer RNA (tRNA). Each tRNA has three unpaired bases known as the anticodon that are complementary to the codon it reads; the tRNA is also covalently attached to the amino acid specified by the complementary codon. When the tRNA binds to its complementary codon in an mRNA strand, the ribosome ligates its amino acid cargo to the new polypeptide chain, which is synthesized from amino terminus to carboxyl terminus. During and after its synthesis, the new protein must fold to its active three-dimensional structure before it can carry out its cellular function.

DNA Replication and Inheritance

The growth, development, and reproduction of organisms relies on cell division, or the process by which a single cell divides into two usually identical daughter cells. This requires first making a duplicate copy of every gene in the genome in a process called DNA replication. The copies are made by specialized enzymes known as DNA polymerases, which "read" one strand of the double-helical DNA, known as the template strand, and synthesize a new complementary strand. Because the DNA double helix is held together by base pairing, the sequence of one strand completely specifies the sequence of its complement; hence only one strand needs to be read by the enzyme to produce a faithful copy. The process of DNA replication is semiconservative; that is, the copy of the genome inherited by each daughter cell contains one original and one newly synthesized strand of DNA.

After DNA replication is complete, the cell must physically separate the two copies of the genome and divide into two distinct membrane-bound cells. In prokaryotes - bacteria and archaea - this usually occurs via a relatively simple process called binary fission, in which each circular genome attaches to the cell membrane and is separated into the daughter cells as the membrane invaginates to split the cytoplasm into two membrane-bound portions. Binary fission is extremely fast compared to the rates of cell division in eukaryotes. Eukaryotic cell division is a more complex process known as the cell cycle; DNA replication occurs during a phase of this cycle known as S phase, while the process of segregating chromosomes and splitting the cytoplasm occurs during M phase. In many single-celled eukaryotes such as yeast, reproduction by budding is common, which results in asymmetrical portions of cytoplasm in the two daughter cells.

Molecular Inheritance

The duplication and transmission of genetic material from one generation of cells to the next is the basis for molecular inheritance, and the link between the classical and molecular pictures of genes. Organisms inherit the characteristics of their parents because the cells of the offspring contain copies of the genes in their parents' cells. In asexually reproducing organisms, the offspring will be a genetic copy or clone of the parent organism. In sexually reproducing organisms, a specialized form of cell division called meiosis produces cells called gametes or germ cells that are haploid, or contain only one copy of each gene. The gametes produced by females are called eggs or ova, and those produced by males are called sperm. Two gametes fuse to form a fertilized egg, a single cell that once again has a diploid number of genes - each with one copy from the mother and one copy from the father.

During the process of meiotic cell division, an event called genetic recombination or *crossing-over* can sometimes occur, in which a length of DNA on one chromatid is swapped with a length of DNA on the corresponding sister chromatid. This has no effect if the alleles on the chromatids are the same, but results in reassortment of otherwise linked alleles if they are different. The Mendelian principle of independent assortment asserts that each of a parent's two genes for each trait will sort independently into gametes; which allele an organism inherits for one trait is unrelated to which allele it inherits for another trait. This is in fact only true for genes that do not reside on the same chromosome, or are located very far from one another on the same chromosome. The closer two genes lie on the same chromosome, the more closely they will be associated in gametes and the more often they will appear together; genes that are very close are essentially never separated because it is extremely unlikely that a crossover point will occur between them. This is known as genetic linkage.

Mutation

DNA replication is for the most part extremely accurate, with an error rate per site of around 10-6 to 10-10 in eukaryotes. Rare, spontaneous alterations in the base sequence of a particular gene arise from a number of sources, such as errors in DNA replication and the aftermath of DNA damage. These errors are called mutations. The cell contains many DNA repair mechanisms for preventing mutations and maintaining the integrity of the genome; however, in some cases — such as breaks in both DNA strands of a chromosome — repairing the physical damage to the molecule is a higher priority than producing an exact copy. Due to the degeneracy of the genetic code, some mutations in protein-coding genes are *silent*, or produce no change in the amino acid sequence of the protein for which they code; for example, the codons UCU and UUC both code for serine, so the U"!C mutation has no effect on the protein. Mutations that do have phenotypic effects are most often neutral or deleterious to the organism, but sometimes they confer benefits to the organism's fitness.

Mutations propagated to the next generation lead to variations within a species' population. Variants of a single gene are known as alleles, and differences in alleles may give rise to differences in traits. Although it is rare for the variants in a single gene to have clearly distinguishable phenotypic effects, certain well-defined traits are in fact controlled by single genetic loci. A gene's most common allele is called the wild type allele, and rare alleles are called mutants. However, this does not imply that the wild-type allele is the ancestor from which the mutants are descended.

The Genome

Chromosomal Organization

The total complement of genes in an organism or cell is known as its genome. In prokaryotes, the vast majority of genes are located on a single chromosome of circular DNA, while eukaryotes usually possess multiple individual linear DNA helices packed into dense DNA-protein complexes called chromosomes. Extrachromosomal DNA is present in many prokaryotes and some simple eukaryotes as small, circular pieces of DNA called plasmids, which usually contain only a few genes each. Generally, regulatory regions and junk DNA are considered to be part of an organism's genome, but structural regions such as telomeres are not. The location (or locus) of a gene and the chromosome on which it is situated is in a sense arbitrary. Genes that appear together on the chromosomes of one species, such as humans, may appear on separate chromosomes in another species, such as mice. Two genes positioned near one another on a chromosome may encode proteins that figure in the same cellular process or in completely unrelated processes. As an example of the former, many of the genes involved in spermatogenesis reside together on the Y chromosome.

Many species carry more than one copy of their genome within each of their somatic cells. Cells or organisms with only one copy of each gene are called haploid; those with two copies are called diploid; and those with more than two copies are called polyploid. When more than one copy is present, the two copies are not necessarily identical; in sexually reproducing organisms, one copy is normally inherited from each parent. The copies may contain distinct DNA sequences encoding distinct alleles.

Composition of the Genome

Typical numbers of genes and size of genomes vary widely among organisms, even those that are fairly closely evolutionarily related. Although it was believed before the completion of the Human Genome Project that the human genome would contain many more genes than simpler animals such as mice or fruit flies, the completion of the project has revealed that the human genome has an unexpectedly low gene density. Estimates of the number of genes in a genome are difficult to compile because they depend on gene finding algorithms that search for patterns resembling those present in known genes, such as open reading frames, promoter regions with sequences resembling the consensus promoter sequence, and related regulatory regions such as TATA boxes in eukaryotes. Gene finding is less reliable in eukaryotic than in prokaryotic genomes due to the presence of non-coding DNA such as introns and pseudogenes. Computational gene finding methods are still significantly more reliable than earlier techniques that required mapping the locations of specific mutations that gave rise to distinguishable alleles.

Gene Content and Genome Size of Various Organisms

Species	*Genome size (Mb)*	*Number of genes*
Mycoplasma genitalium	0.58	500
Streptococcus pneumoniae	2.2	2300
Escherichia coli	4.6	4400
Saccharomyces cerevisiae	12	5800
Arabidopsis thaliana	125	25,500
Caenorhabditis elegans	97	19,000
Sea urchin	814	23,300
Drosophila melanogaster	180	13,700
Mus musculus	2500	29,000
Homo sapiens	2900	27,000
Oryza sativa	466	45-55,000

In most eukaryotic species, very little of the DNA in the genome encodes proteins, and the genes may be separated by vast regions of non-coding DNA, much of which has been labeled "junk DNA" due to its apparent lack of function in the modern organism. A commonly studied type of "junk DNA" is the pseudogenes, or region of non-coding DNA that resembles expressed genes but usually lacks appropriate promoters and other control sequences; such regions are hypothesized to be the results of gene duplication events in a lineage's evolutionary past. Moreover, the genes are often fragmented internally by non-coding sequences called introns, which can be many times longer than the coding sequence but are spliced during post-transcriptional modification of pre-mRNA.

Genetic and Genomic Nomenclature

Gene nomenclature has been established by the HUGO Gene Nomenclature Committee (HGNC) for each known human gene in the form of an approved gene name and symbol (short-form abbreviation). All approved symbols are stored in the HGNC Database. Each symbol is unique and each gene is only given one approved gene symbol. It is necessary to provide a unique symbol for each gene so that people can talk about them. This also facilitates electronic data retrieval from publications. In preference each symbol maintains parallel construction in different members of a gene family and can be used in other species, especially the mouse.

Evolutionary Concept of a Gene

George C. Williams first explicitly advocated the gene-centric view of evolution in his 1966 book *Adaptation and Natural Selection*. He proposed an evolutionary concept of gene to be used when we are talking about natural selection favoring some genes. The definition is: "that which segregates and recombines with appreciable frequency." According to this definition, even an asexual genome could be considered a gene, insofar it have an appreciable permanency through many generations.

The difference is: the molecular gene *transcribes* as a unit, and the evolutionary gene *inherits* as a unit.

Richard Dawkins' *The Selfish Gene* and *The Extended Phenotype* defended the idea that the gene is the only replicator in living systems. This means that only genes transmit their structure largely intact and are potentially immortal in the form of copies. So, genes should be the unit of selection. In *The Selfish Gene* Dawkins attempts to redefine the word 'gene' to mean "an inheritable unit" instead of the generally accepted definition of "a section of DNA coding for a particular protein". In *River Out of Eden*, Dawkins further refined the idea of gene-centric selection by describing life as a river of compatible genes flowing through geological time. Scoop up a bucket of genes from the river of genes, and we have an organism serving as temporary bodies or survival machines. A river of genes may fork into two branches representing two non-interbreeding species as a result of geographical separation.

The *Gene* Concept is Still Changing

The concept of the gene has changed considerably. Originally considered a "unit of inheritance" to a usually DNA-based unit that can exert its effects on the organism through RNA or protein products. It was also previously believed that one gene makes one protein; this concept has been overthrown by the discovery of alternative splicing and trans-splicing.

And the definition of gene is still changing. The first cases of RNA-based inheritance have been discovered in mammals. In plants, cases of traits reappearing after several generations of absence have lead researchers to hypothesise RNA-directed overwriting of genomic DNA. Evidence is also accumulating that the control regions of a gene do not necessarily have to be close to the coding sequence on the linear molecule or even on the same chromosome. Spilianakis and colleagues discovered that the promoter region of the interferon-gamma gene on chromosome 10 and the regulatory regions of the T(H)2 cytokine locus on chromosome 11 come into close proximity in the nucleus possibly to be jointly regulated.

The concept that genes are clearly delimited is also being eroded. There is evidence for fused proteins stemming from two adjacent genes that can produce two separate protein products. While it is not clear whether these fusion proteins are functional, the phenomena is more frequent than previously thought. Even more ground-breaking than the discovery of fused genes is the observation that some proteins can be composed of exons from far away regions and even different chromosomes. This new data has led to an updated, and probably tentative, definition of a gene as "a union of genomic sequences encoding a coherent set of potentially overlapping functional products." This new definition categorizes genes by functional products, whether they be proteins or RNA, rather than specific DNA loci; all regulatory elements of DNA are therefore classified as *gene-associated* regions.

TRANSGENE

A transgene is a gene or genetic material which has been transferred by any of a number of genetic engineering techniques from one organism to another.

In its most precise usage, the term transgene describes a segment of DNA containing a gene sequence which has been isolated from one organism and is introduced into a different organism. This non-native segment of DNA may retain the ability to produce RNA or protein in the transgenic organism or it may alter the normal function of the transgenic organism's genetic code. Typically the DNA is incorporated into the organism's germ line. For example, in higher vertebrates this can be accomplished by injecting the foreign DNA into the nucleus of a fertilized ovum. This technique is routinely used to introduce human disease genes or other genes of interest into strains of laboratory mice to study the function or pathology involved with that particular gene.

In looser usage, transgene can describe any DNA sequence, regardless of whether it contains a gene coding sequence or has been artificially constructed, which has been introduced into an organism or vector construct in which it was previously unknown.

In practical terms, a transgene can be either based on a cDNA (complementary DNA), which is a copy of mRNA (messenger RNA) or on the gene itself residing in its original region of genomic DNA. The difference between these two lies with the facts that the cDNA has been processed to remove introns and also, usually, does not include the regulatory signals that are embedded around and in the gene. The advent of annotated cloned regions of the genome alongside the genome sequence, particularly as large clones in BACs (bacterial artificial chromosomes) or fosmids, and recombineering, which is the method that permits the engineering of these large clones, has changed the practice of transgenesis from its origins with cDNA-based constructs towards the more reliable genomic-based constructs.

PARATRANSGENESIS

Paratransgenesis is a technique that attempts to eliminate a pathogen from vector populations through transgenesis of a symbiont of the vector. The goal of this technique is to control vector-borne diseases. The first step is to identify proteins that prevent the vector species from transmitting the pathogen. The genes coding for these proteins are then introduced into the symbiont, so that they can be expressed in the vector. The final step in the strategy is to introduce these transgenic symbionts into vector populations in the wild.

The first example of this technique used *Rhodnius prolixus* which is associated with the symbiont *Rhodococcus rhodnii*. *Rhodnius prolixus* is an important insect vector of Chagas's disease that is caused by *Trypanosoma cruzi*. The strategy was to engineer *R. rhodnii* to express proteins such as Cecropin A that are toxic to *Trypanosoma cruzi* or that block the transmission of *Trypanosoma cruzi*.

In order to perform paratransgenesis, there are several requirements:

- The Symbiotic bacteria can be grown *in vitro* easily.
- They can be genetically modified, such as through transformation with a plasmid containing the desired gene.
- The engineered symbiont is stable and safe.
- The association between vector and symbiont cannot be attenuated.
- Field delivery is easily handled.

EUGENICS

Eugenics is a social philosophy which advocates the improvement of human hereditary traits through various forms of intervention. Throughout history, eugenics has been regarded by its various advocates as a social responsibility, an altruistic stance of a society, meant to create healthier and more intelligent people, to save resources, and lessen human suffering. More controversially, some, such as the Nazi regime in Germany, used eugenics as a pretext for racial discrimination.

Earlier proposed means of achieving these goals focused on selective breeding, while modern ones focus on prenatal testing and screening, genetic counseling, birth control, in vitro fertilization, and genetic engineering. Opponents argue that eugenics is immoral and is based on, or is itself, pseudoscience.

Historically, eugenics has been used as a justification for coercive state-sponsored discrimination and human rights violations, such as forced sterilization of persons who are claimed to have genetic defects, the killing of the institutionalized population and, in some cases, outright genocide of races perceived as inferior or undesirable.

The modern field and term were first formulated by Sir Francis Galton in 1865, drawing on the recent work of his cousin Charles Darwin. From its inception eugenics was supported by prominent people, including Alexander Graham Bell, George Bernard Shaw, and Margaret Sanger. G. K. Chesterton was against the philosophy of eugenics and he expresses this in his book, *Eugenics and Other Evils*. Eugenics became an academic discipline at many colleges and universities. Funding was provided by prestigious sources such as the Rockefeller Foundation, the Carnegie Institution of Washington, and the Harriman family. Three International Eugenics Conferences presented a global venue for eugenicists with meetings in 1912 in London, and in 1921 and 1932 in New York. Eugenics' scientific reputation started to tumble in the 1930s, a time when Ernst Rüdin began incorporating eugenic rhetoric into the racial policies of Nazi Germany.

Since the postwar period, both the public and the scientific communities have associated eugenics with Nazi abuses, such as enforced racial hygiene, human experimentation, and the extermination of undesired population groups. However, developments in genetic, genomic, and reproductive technologies at the end of the 20th century have raised many new questions and concerns about what exactly constitutes the meaning of *eugenics* and what its ethical and moral status is in the modern era.

Meanings and Types of Eugenics

The word *eugenics* etymologically derives from the Greek word *eu* (*good* or *well*) and the suffix *-gençs* (*born*), and was coined by Sir Francis Galton in 1883.

Eugenics has, from the very beginning, meant many different things to many different people. Historically, the term has referred to everything from prenatal care for mothers to forced sterilization and euthanasia. Much debate took place in the past, and takes place today, as to what exactly counts as eugenics. Some types of eugenics, such as race-based eugenics and class-based eugenics, are sometimes called 'pseudo-eugenics' by proponents of strict eugenics that deals only with perceived beneficial and detrimental genetic traits.

The term *eugenics* is often used to refer to movements and social policies that were influential during the early 20th century. In a historical and broader sense, eugenics can also be a study of "improving human genetic qualities". It is sometimes broadly applied to describe any human action whose goal is to improve the gene pool. Some forms of infanticide in ancient societies, present-day reprogenetics, preemptive abortions and designer babies have been (sometimes controversially) referred to as eugenic.

Because of its normative goals and historical association with scientific racism, as well as the development of the science of genetics, the western scientific community has mostly disassociated itself from the term "eugenics", although one can find advocates of what is now known as *liberal eugenics*.

Ideological social determinists, some of which have obtained college degrees in fields relevant to eugenics, often describe eugenics as a pseudoscience. Modern inquiries into the potential use of genetic engineering have led to an increased invocation of the history of eugenics in discussions of bioethics, most often as a cautionary tale. Some ethicists suggest that even non-coercive eugenics programs would be inherently unethical, though this view has been challenged by such thinkers as Nicholas Agar.

Eugenicists advocate specific policies that (if successful) would lead to a perceived improvement of the human gene pool. Since defining what improvements are desired or beneficial is perceived by many as a cultural choice rather than a matter that can be determined objectively (e.g., by empirical, scientific inquiry), eugenics has often been deemed a pseudoscience. The most disputed aspect of eugenics has been the definition of "improvement" of the human gene pool, such as what is a beneficial characteristic and what is a defect. This aspect of eugenics has historically been tainted with scientific racism.

Early eugenicists were mostly concerned with perceived intelligence factors that often correlated strongly with social class. Many eugenicists took inspiration from the selective breeding of animals (where purebreds are often strived for) as their analogy for improving human society. The mixing of races (or miscegenation) was usually considered as something to be avoided in the name of racial purity. At the time this concept appeared to have some scientific support, and it remained a contentious issue until the advanced development of genetics led to a scientific consensus that the division of the human species into unequal races is unjustifiable. Some see this as an ideological consensus, since equality, just like inequality, is a cultural choice rather than a matter that can be determined objectively.

Eugenics has also been concerned with the elimination of hereditary diseases such as hemophilia and Huntington's disease. However, there are several problems with labeling certain factors as "genetic defects":

- In many cases there is no scientific consensus on what a "genetic defect" is. It is often argued that this is more a matter of social or individual choice.
- What appears to be a "genetic defect" in one context or environment may not be so in another. This can be the case for genes with a heterozygote advantage, such as sickle cell anemia or Tay-Sachs disease, which in their heterozygote form may offer an advantage against, respectively, malaria and tuberculosis.

- Although some birth defects are uniformly lethal, disabled persons can succeed in life.
- Many of the conditions early eugenicists identified as inheritable (pellagra is one such example) are currently considered to be at least partially, if not wholly, attributed to environmental conditions.

Similar concerns have been raised when a prenatal diagnosis of a congenital disorder leads to abortion (see also preimplantation genetic diagnosis).

Eugenic policies have been conceptually divided into two categories:

Positive eugenics is aimed to encourage reproduction among the genetically advantaged. Possible approaches include financial and political stimuli, targeted demographic analyses, in vitro fertilization, egg transplants, and cloning.

Negative eugenics is aimed at lowering fertility among the genetically disadvantaged. This includes abortions, sterilization, and other methods of family planning.

Both positive and negative eugenics can be coercive. Abortion by "fit" women was illegal in Nazi Germany and in the Soviet Union between 1936 and 1968.

During the 20th century, many countries enacted various eugenics policies and programs, including:

- Genetic screening
- Birth control
- Promoting differential birth rates
- Marriage restrictions
- Segregation (both racial segregation as well as segregation of the mentally ill from the normal)
- Compulsory sterilization
- Forced abortions, or, conversely, forced pregnancies
- Genocide

Most of these policies were later regarded as coercive, restrictive, or genocidal, and now few jurisdictions implement policies that are explicitly labeled as eugenic or unequivocally eugenic in substance (however labeled). However, some private organizations assist people in genetic counseling, and reprogenetics may be considered as a form of non-state-enforced "liberal" eugenics.

There are 3 main ways by which the methods of eugenics can be applied. They are:

- mandatory eugenics, in which the government mandates a eugenics program.
- promotional voluntary eugenics, in which eugenics is voluntarily practiced and promoted to the general population, but not officially mandated.
- private eugenics, which is practiced voluntarily by individuals and groups, but not promoted to the general population.

History

Pre-Galtonian Eugenic Philosophies

The basic ideals of eugenics can be found from the beginnings of Western civilization. The philosophy was most famously expounded by Plato, who believed human reproduction should be monitored and controlled by the state. The basic eugenic principle from Plato's The Republic was, "The best men must have intercourse with the best women as frequently as possible, and the opposite is true of the very inferior." Plato understood that this form of government control would not be readily accepted, and proposed that the truth be concealed from the public via a fixed lottery. Mates, in Plato's Republic, would be chosen by a "marriage number" in which the quality of the individual would be quantitatively analyzed, and persons of high numbers would be allowed to procreate with other persons of high numbers. In theory, this would lead to predictable results and the improvement of the human race. However, Plato acknowledged the failure of the "marriage number" since "gold soul" persons could still produce "bronze soul" children. This might have been one of the earliest attempts to mathematically analyze genetic inheritance, which was not perfected until the development of Mendelian genetics and the mapping of the human genome. Other ancient civilizations, such as Rome and Sparta, practiced infanticide as a form of phenotypic selection. In Sparta, newborns were inspected by the city's elders, who decided the fate of the infant. This would be done through abandonment of "weak" or undesirable babies on the slopes of Mount Taygetos, and trials for babies which included bathing them in wine and exposing them to the elements. To Sparta, this would ensure that only the strongest survived and procreated.

The 12 Tables of Roman Law, established early in the formation of the Roman Republic, stated in the fourth table that deformed children would be put to death. In addition, patriarchs in Roman society were given the right to "discard" infants at their discretion. This was often done by drowning undesired newborns in the Tiber River. The practice of infanticide in the ancient world did not subside until the Christianization of the Roman empire.

Galton's Theory

During the 1860s and 1870s, Sir Francis Galton systematized these ideas and practices according to new knowledge about the evolution of man and animals provided by the theory of his cousin Charles Darwin. After reading Darwin's *Origin of Species*, Galton built upon Darwin's ideas whereby the mechanisms of natural selection were potentially thwarted by human civilization. He reasoned that, since many human societies sought to protect the underprivileged and weak, those societies were at odds with the natural selection responsible for extinction of the weakest. Only by changing these social policies, Galton thought, could society be saved from a "reversion towards mediocrity", a phrase that he first coined in statistics and which later changed to the now common "regression towards the mean".

Galton first sketched out his theory in the 1865 article "Hereditary Talent and Character", then elaborated it further in his 1869 book *Hereditary Genius*. He began by studying the way in which human intellectual, moral, and personality traits tended to run in families. Galton's basic argument was that "genius" and "talent" were hereditary traits in humans (although neither he nor Darwin yet had a working model of this type of heredity). He concluded that, since one could use artificial selection to exaggerate traits in other animals, one could expect similar results when applying such models to humans. As he wrote in the introduction to *Hereditary Genius*:

> "I propose to show in this book that a man's natural abilities are derived by inheritance, under exactly the same limitations as are the form and physical features of the whole organic world. Consequently, as it is easy, notwithstanding those limitations, to obtain by careful selection a permanent breed of dogs or horses gifted with peculiar powers of running, or of doing anything else, so it would be quite practicable to produce a highly-gifted race of men by judicious marriages during several consecutive generations."

According to Galton, society already encouraged dysgenic conditions, claiming that the less intelligent were out-reproducing the more intelligent. Galton did not propose any selection methods; rather, he hoped that a solution would be found if social mores changed in a way that encouraged people to see the importance of breeding.

Galton first used the word *eugenic* in his 1883 *Inquiries into Human Faculty and Its Development*, a book in which he meant "to touch on various topics more or less connected with that of the cultivation of race, or, as we might call it, with 'eugenic' questions." He included a footnote to the word "eugenic" which read:

> That is, with questions bearing on what is termed in Greek, *eugenes* namely, good in stock, hereditarily endowed with noble qualities. This, and the allied

words, *eugeneia*, etc., are equally applicable to men, brutes, and plants. We greatly want a brief word to express the science of improving stock, which is by no means confined to questions of judicious mating, but which, especially in the case of man, takes cognisance of all influences that tend in however remote a degree to give to the more suitable races or strains of blood a better chance of prevailing speedily over the less suitable than they otherwise would have had. The word *eugenics* would sufficiently express the idea; it is at least a neater word and a more generalised one than *viriculture* which I once ventured to use.

In 1904 he clarified his definition of eugenics as "the science which deals with all influences that improve the inborn qualities of a race; also with those that develop them to the utmost advantage."

Galton's formulation of eugenics was based on a strong statistical approach, influenced heavily by Adolphe Quetelet's "social physics". Unlike Quetelet, however, Galton did not exalt the "average man" but decried him as mediocre. Galton and his statistical heir Karl Pearson developed what was called the biometrical approach to eugenics, which developed new and complex statistical models (later exported to wholly different fields) to describe the heredity of traits. However, with the rediscovery of Gregor Mendel's hereditary laws, two separate camps of eugenics advocates emerged. One was made up of statisticians, the other of biologists. Statisticians thought the biologists had exceptionally crude mathematical models, while biologists thought the statisticians knew little about biology.

Eugenics eventually referred to human selective reproduction with an intent to create children with desirable traits, generally through the approach of influencing differential birth rates. These policies were mostly divided into two categories: *positive eugenics*, the increased reproduction of those seen to have advantageous hereditary traits; and *negative eugenics*, the discouragement of reproduction by those with hereditary traits perceived as poor. Negative eugenic policies in the past have ranged from attempts at segregation to sterilization and even genocide. Positive eugenic policies have typically taken the form of awards or bonuses for "fit" parents who have another child. Relatively innocuous practices like marriage counseling had early links with eugenic ideology.

Eugenics differed from what would later be known as Social Darwinism. While both claimed intelligence was hereditary, eugenics asserted that new policies were needed to actively change the status quo towards a more "eugenic" state, while the Social Darwinists argued society itself would naturally "check" the problem of "dysgenics" if no welfare policies were in place (for example, the poor might reproduce more but would have higher mortality rates).

Sparta

In Sparta, eugenics was practised, not scientifically but selectively and judged by the infant's ability to live. If the child was deemed incapable of living, it was usually thrown from the Taygetus mountain. It was more common for girls than boys to be killed this way.

Adolf Hitler considered Sparta to be the first "Völkisch State," and much like Ernst Haeckel before him, praised Sparta due to its primitive form of eugenics practice of selective infanticide policy which was applied on deformed children.

Nazi Germany

Nazi Germany under Adolf Hitler was infamous for eugenics programs which attempted to maintain a "pure" German race through a series of programs that ran under the banner of "racial hygiene". Among other activities, the Nazis performed extensive experimentation on live human beings to test their genetic theories, ranging from simple measurement of physical characteristics to the horrific experiments carried out by Josef Mengele for Otmar von Verschuer on twins in the concentration camps. During the 1930s and 1940s, the Nazi regime forcibly sterilized hundreds of thousands of people whom they viewed as mentally and physically "unfit", an estimated 400,000 between 1934 and 1937. The scale of the Nazi program prompted American eugenics advocates to seek an expansion of their program, with one complaining that "the Germans are beating us at our own game". The Nazis went further, however, killing tens of thousands of the institutionalized disabled through compulsory "euthanasia" programs.

They also implemented a number of "positive" eugenics policies, giving awards to "Aryan" women who had large numbers of children and encouraged a service in which "racially pure" single women could deliver illegitimate children. Allegations that such women were also impregnated by SS officers in the *Lebensborn* are common, but unproven. Also, "racially valuable" children from occupied countries were forcibly removed from their parents and adopted by German people. Many of their concerns for eugenics and racial hygiene were also explicitly present in their systematic killing of millions of "undesirable" people including Jews, Gypsies, Jehovah's Witnesses and homosexuals during the Holocaust (much of the killing equipment and methods employed in the death camps were first developed in the euthanasia program). The scope and coercion involved in the German eugenics programs along with a strong use of the rhetoric of eugenics and so-called "racial science" throughout the regime created an indelible cultural association between eugenics and the Third Reich in the postwar years.

The alleged relation between eugenics and the Jewish Holocaust is also questioned.

Contrary to popular beliefs Hitler did not regard the Jews as intellectually inferior and did not send them to the concentration camps on these grounds. In fact, in the 1930s Germans regarded the Jews as a highly talented people. Hitler had different reasons for his genocidal policies toward the Jews. Seymour W. Itzkoff writes that the Holocaust was "a vast dysgenic program to rid Europe of highly intelligent challengers to the existing Christian domination by a numerically and politically minuscule minority". Therefore, according to Itzkoff, "the Holocaust was the very antithesis of eugenic practice."

Eugenics and the United States, 1890s–1945

One of the earliest modern advocates of eugenics (before it was labeled as such) was Alexander Graham Bell. In 1881, Bell investigated the rate of deafness on Martha's Vineyard, Massachusetts. From this he concluded that deafness was hereditary in nature and recommended a marriage prohibition against the deaf ("Memoir upon the formation of a deaf variety of the human Race") despite his own marriage to a deaf woman. Like many other early eugenicists, Bell proposed controlling immigration for the purpose of eugenics, and warned that boarding schools for the deaf could possibly be considered as breeding places of a deaf human race.

In 1907, Indiana became the first of more than thirty states to adopt legislation aimed at compulsory sterilization of certain individuals. Although the law was overturned by the Indiana Supreme Court in, the U.S. Supreme Court upheld the constitutionality of a Virginia law allowing for the compulsory sterilization of patients of state mental institutions in 1927.

Beginning with Connecticut in 1896, many states enacted marriage laws with eugenic criteria, prohibiting anyone who was "epileptic, imbecile or feeble-minded" from marrying. In 1898 Charles B. Davenport, a prominent American biologist, began as director of a biological research station based in Cold Spring Harbor where he experimented with evolution in plants and animals. In 1904 Davenport received funds from the Carnegie Institution to found the Station for Experimental Evolution. The Eugenics Record Office opened in 1910 while Davenport and Harry H. Laughlin began to promote eugenics.

During the 20th century, researchers became interested in the idea that mental illness could run in families and conducted a number of studies to document the heritability of such illnesses as schizophrenia, bipolar disorder, and depression. Their findings were used by the eugenics movement as proof for its cause. State laws were written in the late 1800s and early 1900s to prohibit marriage and force sterilization of the mentally ill in order to prevent the "passing on" of mental illness to the next generation. These laws were upheld by the U.S. Supreme Court in 1927 and were not abolished until the mid-20th century. By 1945 over 45,000 mentally ill individuals in the United States had been forcibly sterilized. All in all, 60,000 Americans suffered from sterilization.

In years to come, the ERO collected a mass of family pedigrees and concluded that those who were unfit came from economically and socially poor backgrounds. Eugenicists such as Davenport, the psychologist Henry H. Goddard and the conservationist Madison Grant (all well respected in their time) began to lobby for various solutions to the problem of the "unfit". (Davenport favored immigration restriction and sterilization as primary methods; Goddard favored segregation in his *The Kallikak Family*; Grant favored all of the above and more, even entertaining the idea of extermination.) Though their methodology and research methods are now understood as highly flawed, at the time this was seen as legitimate scientific research. It did, however, have scientific detractors (notably, Thomas Hunt Morgan, one of the few Mendelians to explicitly criticize eugenics), though most of these focused more on what they considered the crude methodology of eugenicists, and the characterization of almost every human characteristic as being hereditary, rather than the idea of eugenics itself.

Some states sterilized "imbeciles" for much of the 20th century. The U.S. Supreme Court ruled in the 1927 *Buck v. Bell* case that the state of Virginia could sterilize those it thought unfit. The most significant era of eugenic sterilization was between 1907 and 1963, when over 64,000 individuals were forcibly sterilized under eugenic legislation in the United States. A favorable report on the results of sterilization in California, the state with the most sterilizations by far, was published in book form by the biologist Paul Popenoe and was widely cited by the Nazi government as evidence that wide-reaching sterilization programs were feasible and humane. When Nazi administrators went on trial for war crimes in Nuremberg after World War II, they justified the mass sterilizations (over 450,000 in less than a decade) by citing the United States as their inspiration.

The idea of "genius" and "talent" is also considered by William Graham Sumner, a founder of the American Sociological Society (now called the American Sociological Association). He maintained that if the government did not meddle with the social policy of *laissez-faire*, a class of genius would rise to the top of the system of social stratification, followed by a class of talent. Most of the rest of society would fit into the class of mediocrity. Those who were considered to be defective (mentally retarded, handicapped, etc.) had a negative effect on social progress by draining off necessary resources. They should be left on their own to sink or swim. But those in the class of delinquent (criminals, deviants, etc.) should be eliminated from society ("Folkways", 1907).

Eugenics is today often associated with racism and both W.E.B. DuBois and

Marcus Garvey supported eugenics or ideas resembling eugenics as a way to reduce African American suffering and improve their stature. However, methods of eugenics were applied to reformulate more restrictive definitions of white racial purity in existing state laws banning interracial marriage: the so-called anti-miscegenation laws. The most famous example of the influence of eugenics and its emphasis on strict racial segregation on such "anti-miscegenation" legislation was Virginia's Racial Integrity Act of 1924. The U.S. Supreme Court overturned this law in 1967 in Loving v. Virginia, and declared anti-miscegenation laws unconstitutional.

With the passage of the Immigration Act of 1924, eugenicists for the first time played a central role in the Congressional debate as expert advisers on the threat of "inferior stock" from eastern and southern Europe. This reduced the number of immigrants from abroad to 15 percent from previous years, to control the number of "unfit" individuals entering the country. The new act, inspired by the eugenic belief in the racial superiority of "old stock" white Americans as members of the "Nordic race" (a form of white supremacy), strengthened the position of existing laws prohibiting race- mixing. Eugenic considerations also lay behind the adoption of incest laws in much of the U.S. and were used to justify many anti-miscegenation laws.

Various authors, notably Stephen Jay Gould, have repeatedly asserted that restrictions on immigration passed in the United States during the 1920s (and overhauled in 1965 with the Immigration and Nationality Act) were motivated by the goals of eugenics, in particular, a desire to exclude races considered to be inferior from the national gene pool. During the early 20th century, the United States and Canada began to receive far higher numbers of Southern and Eastern European immigrants. Influential eugenicists like Lothrop Stoddard and Harry Laughlin (who was appointed as an expert witness for the House Committee on Immigration and Naturalization in 1920) presented arguments that these were inferior races that would pollute the national gene pool if their numbers went unrestricted. It has been argued that this stirred both Canada and the United States into passing laws creating a hierarchy of nationalities, rating them from the most desirable Anglo-Saxon and Nordic peoples to the Chinese and Japanese immigrants, who were almost completely banned from entering the country. However, several people, in particular Franz Samelson, Mark Snyderman and Richard Herrnstein, have argued that, based on their examination of the records of the congressional debates over immigration policy, Congress gave virtually no consideration to these factors. According to these authors, the restrictions were motivated primarily by a desire to maintain the country's cultural integrity against a heavy influx of foreigners. This interpretation is not, however, accepted by most historians of eugenics.

Some who disagree with the idea of eugenics in general contend that eugenics legislation still had benefits. Margaret Sanger (founder of Planned Parenthood of America) found it a useful tool to urge the legalization of contraception. In its time eugenics was seen by many as scientific and progressive, the natural application of knowledge about breeding to the arena of human life. Before the death camps of World War II, the idea that eugenics could lead to genocide was not taken seriously.

Japan

In the early part of the Shôwa era, Japanese governments executed a eugenic policy to limit the birth of children with "inferior" traits, as well as aiming to protect the life and health of mothers.

The *Race Eugenic Protection Law* was submitted from 1934 to 1938 to the Diet. After four amendments, this draft was promulgated as the *National Eugenic Law* in 1940 by the Konoe government. According to the *Eugenic Protection Law* (1948), sterilization could be enforced on criminals "with genetic predisposition to commit crime", patients with genetic diseases such as total color-blindness, hemophilia, albinism and ichthyosis, and mental affections such as schizophrenia, manic-depressiveness and epilepsy.. Mental illnesses were added in 1952.

Canada

In Canada, the eugenics movement took place early in the 20th Century, particularly in Alberta, and was quite popular. The Sexual Sterilization Act of Alberta was enacted in 1928, focusing the movement on the sterilization of mentally deficient individuals, as determined by the Alberta Eugenics Board. The campaign to enforce this action was backed by groups such as the United Farm Women's Group, including key member Emily Murphy.

Individuals were assessed using IQ tests like the Stanford-Binet. This posed a problem to new immigrants arriving in Canada, as many had not mastered the English language, and often their scores denoted them as having impaired intellectual functioning. As a result, many of those sterilized under the Sexual Sterilization Act were immigrants who were unfairly categorized.

The popularity of the eugenics movement peaked during the depression. Individuals sought an explanation for the financial problems of the nation, and the notion of defective breeding became a scapegoat; citizens blamed individuals considered to be subhuman. The end of the Canadian eugenics movement was brought about when the Sexual Sterilization Act was repealed in 1972.

Australia

The policy of removing Aboriginal children from their parents emerged from an opinion based on Eugenics theory in late nineteenth and early twentieth-century Australia that the 'full-blood' tribal Aborigine would be unable to sustain itself, and was doomed to inevitable extinction. An ideology at the time held that mankind could be divided into a civilisational hierarchy. This white supremacist notion supposed that Northern Europeans were superior in civilisation and that Aborigines were inferior. According to this view, the increasing numbers of mixed-descent children in Australia, labelled as 'half-castes' (or alternatively 'crossbreeds', 'quadroons' and 'octoroons'), were widely seen to be a threat to racial purity. In the first half of the twentieth century, this led to policies and legislation that resulted in the removal of children from their parents. The stated aim was to culturally assimilate mixed-descent people into contemporary Australian society. In all states and territories legislation was passed in the early years of the twentieth century which gave Aboriginal protectors guardianship rights over Aborigines up to the age of sixteen or twenty-one. Policemen or other agents of the state (such as Aboriginal Protection Officers), were given the power to locate and transfer babies and children of mixed descent, from their mothers or families or communities into institutions. In these Australian states and territories, half-caste institutions (both government or missionary) were established in the early decades of the twentieth-century for the reception of these separated children. The 2002 movie *Rabbit-Proof Fence* portrays this system and the harrowing consequences of attempting to overcome it.

In 1915 A.O. Neville was appointed the second Western Australia State *Chief Protector of Aborigines*. During the next quarter-century, he presided over the now notorious 'Assimilation' policy of removing mixed-race Aboriginal children from their parents. This policy in turn created the Stolen Generations and set in motion a grieving process that through the now widely accepted concept of trans-generational grief, would affect many generations to come. In 1936 Neville became the *Commissioner for Native Affairs*, a post he held until his retirement in.

Neville believed that biological absorption was the key to 'uplifting the Native race.' Speaking before the Moseley Royal Commission, which investigated the administration of Aboriginals in 1934, he defended the policies of forced settlement, removing children from parents, surveillance, discipline and punishment, arguing that "they have to be protected against themselves whether they like it or not. They cannot remain as they are. The sore spot requires the application of the surgeon's knife for the good of the patient, and probably against the patients will."

In his twilight years Neville continued to actively promote his policy. Towards the end of his career, Neville published *Australia's Coloured Minority*, a text outlining his plan for the biological absorption of aboriginal people into white Australia. It is a classic example of the eugenics policies popular at the time in the Western world.

Sweden

From just prior to World War II until 1975, Sweden forcibly sterilized more than 62,000 people with Herman Lundborg in the lead of the project. Sweden's large-scale eugenics program targeted ethnic or racial minorities, and the mentally ill. As was the case in other programs, ethnicity and race were believed to be connected to mental and physical health. While many Swedes disliked the program, politicians generally supported it; the ruling left supported it more as a means of promoting social health, while amongst the right it was more about racial protectionism. (Not until 1999 did the Swedish government begin paying compensation to the victims and their families.)

Britain

In Britain, eugenics never received significant state funding. Furthermore, its emphasis was more upon class, rather than race. Indeed, Galton expressed these views during a lecture in 1901 in which he placed the British society into groups. These groupings are shown in the figure and indicate the proportion of society falling into each group and their perceived genetic worth. Galton suggested that negative eugenics (i.e. an attempt to prevent them from bearing offspring) should be applied only to those in the lowest social group (the "Undesirables"), while positive eugenics applied to the higher classes. However, he appreciated the worth of the higher working classes to society and industry.

Sterilisation programmes were never legalised, although some were carried out in private upon the mentally ill by clinicians who were in favour of a more widespread eugenics plan. (Sterilisation had, in fact, been carried out to prevent masturbation in mentally ill patients since the 1820s, long before the eugenics movement.) Indeed, those in support of eugenics shifted their lobbying of Parliament from enforced to voluntary sterilization, in the hope of achieving more legal recognition. But leave for the Labour MP Major A. G. Church,to propse a Private Member's Bill in 1931, which would legalise the operation for voluntary sterilization, was rejected by 167 votes to 89.

The popularity of eugenics in Britain was reflected by the fact that only two universities established courses in this field (University College London and Liverpool University), but the position of a professorship in eugenics was never created at either. The Galton Institute, affiliated to UCL, was headed by Galton's protégé, Karl Pearson.

Other Countries

Almost all non-Catholic Western nations adopted some eugenic legislations. In July 1933 Germany passed a law allowing for the involuntary sterilization of "hereditary and incurable drunkards, sexual criminals, lunatics, and those suffering from an incurable disease which would be passed on to their offspring." Two provinces in Canada carried out thousands of compulsory sterilizations, and these lasted into the 1970s. Many First Nations (native Canadians) were targeted, as well as immigrants from Eastern Europe, as the program identified racial and ethnic minorities to be genetically inferior[*citation needed*]. Besides the large-scale program in the United States, other nations included Australia, Norway, France, Finland, Denmark, Estonia, Iceland, and Switzerland with programs to sterilize people the government declared to be mentally deficient. Singapore practiced a limited form of eugenics that involved discouraging marriage between university graduates and the rest through segregation in matchmaking agencies, in the hope that the former would produce better children.

Marginalization after World War II

After the experience of Nazi Germany, many ideas about "racial hygiene" and "unfit" members of society were publicly renounced by politicians and members of the scientific community. The Nuremberg Trials against former Nazi leaders revealed to the world many of the regime's genocidal practices and resulted in formalized policies of medical ethics and the 1950 UNESCO statement on race. Many scientific societies released their own similar "race statements" over the years, and the Universal Declaration of Human Rights, developed in response to abuses during the Second World War, was adopted by the United Nations in 1948 and affirmed, "Men and women of full age, without any limitation due to race, nationality or religion, have the right to marry and to found a family." In continuation, the 1978 UNESCO declaration on race and racial prejudice states that the fundamental equality of all human beings is the ideal toward which ethics and science should converge.

In reaction to Nazi abuses, eugenics became almost universally reviled in many of the nations where it had once been popular (however, some eugenics programs, including sterilization, continued quietly for decades). Many pre-war eugenicists engaged in what they later labeled "crypto-eugenics", purposefully taking their eugenic beliefs "underground" and becoming respected anthropologists, biologists and geneticists in the postwar world (including Robert Yerkes in the U.S. and Otmar von Verschuer in Germany). Californian eugenicist Paul Popenoe founded marriage counseling during the 1950s, a career change which grew from his eugenic interests in promoting "healthy marriages" between "fit" couples.

The American Life League, an opponent of abortion, charges that eugenics was merely "re-packaged" after the war, and promoted anew in the guise of the population-control and environmentalism movements. They claim, for example, that Planned Parenthood was funded and cultivated by the Eugenics Society for these reasons. Julian Huxley, the first Director-General of UNESCO and a founder of the World

Wildlife Fund was also a Eugenics Society president and a strong supporte of eugenics

> [E]ven though it is quite true that any radical eugenic policy will be for many years politically and psychologically impossible, it will be important for UNESCO to see that the eugenic problem is examined with the greatest care, and that the public mind is informed of the issues at stake so that much that now is unthinkable may at least become thinkable. —Julian Huxley

High school and college textbooks from the 1920s through the '40s often had chapters touting the scientific progress to be had from applying eugenic principles to the population. Many early scientific journals devoted to heredity in general were run by eugenicists and featured eugenics articles alongside studies of heredity in nonhuman organisms. After eugenics fell out of scientific favor, most references to eugenics were removed from textbooks and subsequent editions of relevant journals. Even the names of some journals changed to reflect new attitudes. For example, *Eugenics Quarterly* became *Social Biology* in 1969 (the journal still exists today, though it looks little like its predecessor). Notable members of the American Eugenics Society (1922–94) during the second half of the 20th century included Joseph Fletcher, originator of Situational ethics; Dr. Clarence Gamble of the Procter & Gamble fortune; and Garrett Hardin, a population control advocate and author of *The Tragedy of the Commons*.

In the United States, the eugenics movement had largely lost most popular and political support by the end of the 1930s while forced sterilizations mostly ended in the 1960s with the last performed in 1981. Many US states continued to prohibit biracial marriages with "anti-miscegenation laws" such as Virginia's The Racial Integrity Act of 1924, until they were over-ruled by the Supreme Court in 1967 in Loving v. Virginia. The Immigration Restriction Act of 1924, which was designed to limit the immigration of "dysgenic" Italians, and eastern European Jews, was repealed and replaced by the Immigration and Nationality Act in 1965.

However, some prominent academics continued to support eugenics after the war. In 1963 the Ciba Foundation convened a conference in London under the title "Man and His Future," at which three distinguished biologists and Nobel laureates (Hermann Muller, Joshua Lederberg, and Francis Crick) all spoke strongly in favor of eugenics.

A few nations, notably, Canada and Sweden, maintained large-scale eugenics programs, including forced sterilization of mentally handicapped individuals, as well as other practices, until the 1970s.

Modern Eugenics, Genetic Engineering, and Ethical Re-valuation

Beginning in the 1980s, the history and concept of eugenics were widely discussed as knowledge about genetics advanced significantly. Endeavors such as the Human Genome Project made the effective modification of the human species seem possible again (as did Darwin's initial theory of evolution in the 1860s, along with the rediscovery of Mendel's laws in the early 20th century). The difference at the beginning of the 21st century was the guarded attitude towards eugenics, which had become a watchword to be feared rather than embraced.

Suggestions and Ideas

A few scientific researchers such as psychologist Richard Lynn, psychologist Raymond Cattell, and doctor Gregory Stock have openly called for eugenic policies using modern technology, but they represent a minority opinion in current scientific and cultural circles. One attempted implementation of a form of eugenics was a "genius sperm bank" (1980–99) created by Robert Klark Graham, from which nearly 230 children were conceived (the best known donor were Nobel Prize winners William Shockley and J.D.Watson): other Laureates including Maurice Wilkins destroyed their invitation letters.). In the U.S. and Europe, though, these attempts have frequently been criticized as in the same spirit of classist and racist forms of eugenics of the 1930s. Because of its association with compulsory sterilization and the racial ideals of the Nazi Party, the word *eugenics* is rarely used by the advocates of such programs.

Eugenicists have argued that immigration from countries with low national IQ is undesirable. According to Raymond Cattell "when a country is opening its doors to immigration from diverse countries, it is like a farmer who buys his seeds from different sources by the sack, with sacks of different average quality of contents."

Cyprus

A similar screening policy (including prenatal screening and abortion) intended to reduce the incidence of thalassemia exists on both sides of the island of Cyprus. Since the program's implementation in the 1970s, it has reduced the ratio of children born with the hereditary blood disease from 1 out of every 158 births to almost zero.

In the government controlled areas, tests for the gene are compulsory for both partners, prior to marriage.

United States

There are some states that require a blood test prior to marriage. While these tests are typically restricted to the detection of the sexually transmitted disease

Syphilis (which was the most common STD at the time these laws were enacted), some partners will voluntarily test for other diseases and genetic incompatibilities.

Harris polls in 1986 and 1992 recorded majority public support for limited forms of germ-line intervention, especially to prevent "children inheriting usually fatal genetic disease".

In 1971, lobbying by the US organisation The International Association for Voluntary Sterilization (AVS), led politicians and officials at the Office for Equal Opportunity to pay for voluntary sterilization of low income Americans for birth-control purposes. AVS also focused on the International community, and its lobbying led to a US foreign policy and funding from the U.S. Agency for International Development to encourage Third World/Developing World countries to utilise abortion and sterilization in order to control their population growth. For further information see EngenderHealth.

Dor Yeshorim

Dor Yeshorim, a program which seeks to reduce the incidence of Tay-Sachs disease, Cystic Fibrosis, Canavan disease, Fanconi anemia, Familial Dysautonomia, Glycogen storage disease, Bloom's Syndrome, Gaucher Disease, Niemann-Pick Disease, and Mucolipidosis IV among certain Jewish communities, is another screening program which has drawn comparisons with liberal eugenics. In Israel, at the expense of the state, the general public is advised to carry out genetic tests to diagnose these diseases before the birth of a baby. If an unborn baby is diagnosed with one of these diseases among which Tay-Sachs is the most commonly known, the pregnancy may be terminated, subject to consent. Most other Ashkenazi Jewish communities also run screening programs because of the higher incidence of genetic diseases. In some Jewish communities, the ancient custom of matchmaking (shidduch) is still practiced, and in order to attempt to prevent the tragedy of infant death which always results from being homozygous for Tay-Sachs, associations such as the strongly observant Dor Yeshorim (which was founded by a rabbi who lost four children to Tay-Sachs with the purpose of preventing others from suffering the same tragedy) test young couples to check whether they carry a risk of passing on fatal conditions. If both the young man and woman are Tay-Sachs carriers, it is common for the match to be broken off. Judaism, like numerous other religions, discourages abortion unless there is a risk to the mother, in which case her needs take precedence. The effort is not aimed at eradicating the hereditary traits, but rather at the occurrence of homozygosity. The actual impact of this program on allele frequencies is unknown, but little impact would be expected because the program does not impose genetic selection. Instead, it encourages disassortative mating.

Ethical Re-Assessment

In modern bioethics literature, the history of eugenics presents many moral and ethical questions. Commentators have suggested the new "eugenics" will come from reproductive technologies that will allow parents to create so-called "designer babies" (what the biologist Lee M. Silver prominently called "reprogenetics"). It has been argued that this "non-coercive" form of biological "improvement" will be predominantly motivated by individual competitiveness and the desire to create "the best opportunities" for children, rather than an urge to improve the species as a whole, which characterized the early 20th-century forms of eugenics. Because of this non-coercive nature, lack of involvement by the state and a difference in goals, some commentators have questioned whether such activities are eugenics or something else altogether. But critics note[*citation needed*] that Francis Galton, did not advocate for coercion when he defined the principles of eugenics. In other words, eugenics does not mean coercion. It is, according to Galton who originated the term, the proper label for bioengineering of "better" human beings.

Daniel Kevles argues that eugenics and the conservation of natural resources are similar propositions. Both can be practiced foolishly so as to abuse individual rights, but both can be practiced wisely.

Some disability activists argue that, although their impairments may cause them pain or discomfort, what really disables them as members of society is a sociocultural system that does not recognize their right to genuinely equal treatment. They express skepticism that any form of eugenics could be to the benefit of the disabled considering their treatment by historical eugenic campaigns.

James D. Watson, the first director of the Human Genome Project, initiated the Ethical, Legal and Social Implications Program (ELSI) which has funded a number of studies into the implications of human genetic engineering (along with a prominent website on the history of eugenics), because:

> In putting ethics so soon into the genome agenda, I was responding to my own personal fear that all too soon critics of the Genome Project would point out that I was a representative of the Cold Spring Harbor Laboratory that once housed the controversial Eugenics Record Office. My not forming a genome ethics program quickly might be falsely used as evidence that I was a closet eugenicist, having as my real long-term purpose the unambiguous identification of genes that lead to social and occupational stratification as well as genes justifying racial discrimination.

Distinguished geneticists including Nobel Prize-winners John Sulston ("I don't think one ought to bring a clearly disabled child into the world") and Watson ("Once you have a way in which you can improve our children, no one can stop it") support genetic screening. Which ideas should be described as "eugenic" are still controversial in both public and scholarly spheres. Some observers such as Philip Kitcher have described the use of genetic screening by parents as making possible a form of "voluntary" eugenics.

Some modern subcultures advocate different forms of eugenics assisted by human cloning and human genetic engineering, sometimes even as part of a new religious movement (see Raëlism, Cosmotheism, or Prometheism). These groups also talk of "neo-eugenics". "conscious evolution", or "genetic freedom".

Behavioral traits often identified as potential targets for modification through human genetic engineering include intelligence, depression, schizophrenia, alcoholism, sexual behavior (and orientation) and criminality.

Criticism

Diseases vs. Traits

While the science of genetics has increasingly provided means by which certain characteristics and conditions can be identified and understood, given the complexity of human genetics, culture, and psychology there is at this point no agreed objective means of determining which traits might be ultimately desirable or undesirable. Eugenic manipulations that reduce the propensity for criminality and violence, for example, might result in the population being enslaved by an outside aggressor it can no longer defend itself against. On the other hand, genetic diseases like hemochromatosis can increase susceptibility to illness, cause physical deformities, and other dysfunctions. Eugenic measures against many of these diseases are already being undertaken in societies around the world, while measures against traits that affect more subtle, poorly understood traits, such as criminality, are relegated to the realm of speculation and science fiction. The effects of diseases are essentially wholly negative, and societies everywhere seek to reduce their impact by various means, some of which are eugenic in all but name. The other traits that are discussed have positive as well as negative effects and are not generally targeted at present anywhere.

Slippery Slope

A common criticism of eugenics is that it inevitably leads to measures that are unethical (Lynn 2001). A hypothetical scenario posits that if one racial minority group is on average less intelligent than the racial majority group, then it is more likely that the racial minority group will be submitted to a eugenics program rather than the least intelligent members of the whole population.

H.L. Kaye wrote of "the obvious truth that eugenics has been discredited by

Hitler's crimes," (Kaye 1989). R. L. Hayman argued "the eugenics movement is an anachronism, its political implications exposed by the Holocaust," (Hayman 1990).

Steven Pinker has stated that it is "a conventional wisdom among left-leaning academics that genes imply genocide." He has responded to this "conventional wisdom" by comparing the history of Marxism, which had the opposite position on genes to that of Nazism:

> But the 20th century suffered "two" ideologies that led to genocides. The other one, Marxism, had no use for race, didn't believe in genes and denied that human nature was a meaningful concept. Clearly, it's not an emphasis on genes or evolution that is dangerous. It's the desire to remake humanity by coercive means (eugenics or social engineering) and the belief that humanity advances through a struggle in which superior groups (race or classes) triumph over inferior ones.

Richard Lynn broadens his criticism of eugenics, by arguing that any social philosophy is capable of ethical misuse. Though Christian principles have aided in the abolition of slavery and the establishment of welfare programs, he notes that the Christian church has also burned many dissidents at the stake and allowed for the killing of large numbers of innocent people by Crusaders. Lynn argues the appropriate response is to condemn these killings, but believes Christianity does not "inevitably [lead] to the extermination of those who do not accept its doctrines," (Lynn 2001).

Genetic Diversity

Eugenic policies could also lead to loss of genetic diversity, in which case a culturally accepted improvement of the gene pool would very likely, as evidenced in numerous instances in isolated island populations (e.g. the Dodo, Raphus cucullatus, of Mauritius) result in extinction due to increased vulnerability to disease, reduced ability to adapt to environmental change and other factors both known and unknown. A long-term eugenics plan might lead to a scenario similar to this because the elimination of traits deemed undesirable would reduce genetic diversity by definition. (Galton 2001, 48).

Proponents of eugenics argue that in any one generation any realistic program would make only minor changes in the gene pool, giving plenty of time to reverse direction if unintended consequences emerge, reducing the likelihood of the elimination of desirable genes. Proponents of eugenics argue that any appreciable reduction in diversity is so far in the future that little concern is needed for now.

The possible elimination of the autism genotype is a significant political issue in the autism rights movement, which claims autism is a form of neurodiversity. Many advocates of Down Syndrome rights also consider Down Syndrome (Trisomy-21) a form of neurodiversity.

Heterozygous Recessive Traits

In some instances efforts to eradicate certain single-gene mutations would be nearly impossible. In the event the condition in question was a heterozygous recessive trait, the problem is that by eliminating the visible unwanted trait, there are still as many genes for the condition left in the gene pool as were eliminated according to the Hardy-Weinberg principle, which states that a population's genetics are defined as pp+2pq+qq at equilibrium. With genetic testing it may be possible to detect all of the heterozygous recessive traits, but only at great cost with the current technology. Under normal circumstances it is only possible to eliminate a dominant allele from the gene pool. Recessive traits can be severely reduced, but never eliminated unless the complete genetic makeup of all members of the pool was known, as aforementioned. As only very few undesirable traits, such as Huntington's disease, are dominant, the practical value for "eliminating" traits is quite low.

However, there are examples of eugenic acts that managed to lower the prevalence of recessive diseases, although not influencing the prevalence of heterozygote carriers of those diseases. The elevated prevalence of certain genetically transmitted diseases among the Ashkenazi Jewish population (Tay-Sachs, Cystic Fibrosis, Canavan's disease and Goucher's disease), has been decreased in current populations by the application of genetic screening.

Counterarguments

Dysgenics

Some supporters of eugenics allege that a dysgenic decline in intelligence is occurring, which may lead to the collapse of civilization, and justify eugenic programs on that basis.

Potential Benefits

Small differences in average IQ at the group level might theoretically have large effects on social outcomes. Richard Herrnstein and Charles Murray altered the mean IQ (100) of the U.S. National Longitudinal Survey of Youth's population sample by randomly deleting individuals below an IQ of 103 until the population mean reached 103. This calculation was conducted twice and averaged together to avoid error from the random selection. This test showed that the new group with an average IQ of 103 had a poverty rate 25% lower than a group with an average IQ of 100. Similar substantial correlations in high school drop-out rates, crime rates, and other outcomes were measured.

Indeed, many studies suggest that IQ correlates with various socioeconomic factors. However, to what extent IQ is a cause of these socioeconomic factors, as opposed to a consequence of them, is disputed. Studies have suggested, for example, that education increases an individual's IQ — although other studies have shown that education has little to no effect.

CLASSICAL GENETICS

Classical genetics consists of the techniques and methodologies of genetics that predate the advent of molecular biology. A key discovery of classical genetics in eukaryotes was genetic linkage. The observation that some genes do not segregate independently at meiosis, broke the laws of Mendelian inheritance, and provided science with a way to map characteristics to a location on the chromosomes. Linkage maps are still used today, especially in breeding for plant improvement.

After the discovery of the genetic code and such tools of cloning as restriction enzymes, the avenues of investigation open to geneticists were greatly broadened. Some classical genetic ideas have been supplanted with the mechanistic understanding brought by molecular discoveries, but many remain intact and in use. Classical genetics is often contrasted with reverse genetics, and aspects of molecular biology are sometimes referred to as molecular genetics.

ECOLOGICAL GENETICS

Ecological genetics is the study of genetics in the context of the interactions among organisms and between the organisms and their environment. While molecular genetics studies the structure and function of genes at a molecular level, ecological genetics (and the related field of population genetics) studies phenotypic evolution in natural populations of organisms. Research in this field is of traits of ecological significance — that is, traits related to fitness, which affect an organism's survival and reproduction (e.g., flowering time, drought tolerance, sex ratio).

Studies are often done on insects and other organisms that have short generation times, and thus evolve at fast rates.

History

Although work on natural populations had been done previously, it is acknowledged that the field was founded by the English biologist E.B. Ford (1901-1988) in the early 20th century. Ford was taught genetics at Oxford University by Julian Huxley, and started research on the genetics of natural populations in 1924. *Ecological Genetics* is the title of his 1964 'magnum opus' on the subject (4th ed 1975). Other notable ecological geneticists would include Theodosius Dobzhansky who worked on chromosome polymorphism in fruit flies. As a young researcher in Russia, Dobzhansky had been influenced by Sergei Chetverikov, who also deserves to be remembered as a founder of genetics in the field, though his significance was not appreciated until much later.

Philip Sheppard, Cyril Clarke, Bernard Kettlewell and A.J. Cain were all strongly influenced by Ford; their careers date from the post WWII era. Collectively, their work on lepidoptera, and on human blood groups, established the field, and threw light on selection in natural populations where its role had been once doubted.

Work of this kind needs long-term funding, as well as grounding in both ecology and genetics. These are both difficult requirements. Research projects can last longer than a researcher's career; for instance, research into mimicry started 150 years ago, and is still going strongly. Funding of this type of research is still rather erratic, but at least the value of working with natural populations in the field cannot now be doubted.

More detail on Ecological genetics may be obtained in Polymorphism (biology).

MOLECULAR GENETICS

Molecular genetics is the field of biology which studies the structure and function of genes at a molecular level. The field studies how the genes are transferred from generation to generation. Molecular genetics employs the methods of genetics and molecular biology. It is so-called to differentiate it from other sub fields of genetics such as ecological genetics and population genetics. An important area within molecular genetics is the use of molecular information to determine the patterns of descent, and therefore the correct scientific classification of organisms: this is called molecular systematics.

Along with determining the pattern of descendants, molecular genetics helps in understanding genetic mutations that can cause certain types of diseases. Through utilizing the methods of genetics and molecular biology, molecular genetics discovers the reasons why traits are carried on and how and why some may mutate.

Forward Genetics

One of the first tools available to molecular geneticists is the forward genetic screen. The aim of this technique is to identify mutations that produce a certain phenotype. A mutagen is very often used to accelerate this process. Once mutants have been isolated, the mutated gene can be molecularly identified.

Reverse Genetics

While forward genetic screens are productive, a more straightforward approach would be to determine the phenotype that results from mutating a given gene. This is called reverse genetics. In some organisms, such as yeast and mice, it is possible to induce the deletion of a particular gene, creating a gene knockout. Alternatives include the random induction of DNA deletions and subsequent selection for deletions in a gene of interest, the application of RNA interference and the creation of transgenic organisms that do not express a gene of interest.

Gene Therapy

Many mutations in a person's genes result in severe medical condition. These mutations cause the protein encoded by that gene to malfunction and cells that rely on this protein cannot function properly. This can cause problems for the tissues or organs. Conditions related to gene mutations are called genetic disorders. One way to fix the internal problems is gene therapy. By adding a corrected copy of the gene, the affected cells, tissues, and organs may work properly. As opposed to drug-based approaches, gene therapy repairs the underlying genetic defect.

Gene Therapy is the process to treat or alleviate diseases by genetically modifying the cells of the infected person, causing the gene to function properly. When a human disease gene has been recognized, molecular genetics tools can be used to explore the process of the gene in both the normal and pathogenic states. From there, the gene is transferred either in vivo or ex vivo and the body begins to identify the new gene. Gene therapy has to be repeated several times for the infected patient to continually be relieved.

Currently, gene therapy is still being experimented with and products are not approved by the U.S. Food and Drug Administration. There have been several set backs in the last 15 years that have restricted further measures on gene therapy. As there are unsuccessful attempts, there continues to be a growing number of successful gene therapy transfers which has furthered the research.

Major diseases that can be treated with gene therapy: infectious diseases (result of infection by a virus or bacterial pathogen), cancers, inherited disorders, immune system disorders

Classical Gene Therapy

Classical gene therapy is the approach which delivers genes to the appropriate target cells with a goal of attaining optimal expression of the new, introduced gene. Once inside the patient, the expressed genes are intended to: produce a product that the patient lacks, kill the diseased cells directly by producing a toxin, and activate the immune system to help the killing of the diseased cells.

Nonclassical Gene Therapy

Nonclassical gene therapy inhibits the expression of genes related to pathogenesis, or corrects a genetic defect and restores normal gene expression.

In vivo Gene Transfer

During In vivo gene transfer the genes are transferred directly into the tissue of the patient. This can be the only possible option in patients with tissues where individual cells cannot be cultured outside the body in sufficient numbers (i.e. brain cells). Also, in vivo gene transfer is necessary when cultured cells cannot be re-implanted in patients effectively.

Ex Vivo Gene Transfer

During Ex vivo gene transfer the cells are cultured outside the body and then the genes are transferred into the cells grown in culture. The cells that have been transformed successfully are expanded by cell culture and then introduced into the patient.

Principles for Gene Transfer

Classical gene therapies usually require efficient transfer of cloned genes into the disease cells to the introduced genes are expressed at high levels. There are several different physicochemical and biological methods that can be used to transfer genes into human cells. The size of the DNA fragments that can be transferred are very limited, and often the transferred gene is not a conventional gene. Horizontal gene transfer is the transfer of genetic material from one cell to another that is not its offspring. Artificial horizontal gene transfer is a form of genetic engineering.

Techniques in Molecular Genetics

There are three general techniques used for molecular genetics: amplification, separation and detection, and expression. Specifically used for amplification is the Polymerase chain reaction, which is an "indispensable tool in a great variety of applications". In the separation and detection technique DNA and mRNA are isolated from their cells. Gene expression in cells or organisms is done in a place or time that is not normal for that specific gene.

Amplification

There are other methods for amplification besides the Polymerase Chain Reaction. Cloning DNA in bacteria is also a way to amplify DNA in genes.

Polymerase Chain Reaction

The main materials used in the Polymerase Chain Reaction are DNA nucleotides, template DNA, primers and Taq polymerase. DNA nucleotides are the base for the new DNA, the template DNA is the specific sequence being amplified, primers are complementary nucleotides that can go on either side of the template DNA, and Taq polymerase is a heat stable enzyme that jump-starts the production of new DNA. This technique does not need to use living bacteria or cells, all that is needed is the base sequence of the DNA needing amplification.

Cloning DNA in Bacteria

The word cloning for this type of amplification entails making multiple identical copies of a sequence of DNA. The target DNA sequence is then inserted into a [cloning vector]. Because this vector originates from a self-replicating virus, plasmid, or higher organism cell when the appropriate size DNA is inserted the "target and vector DNA fragments are then ligated" and create a recombinant DNA molecule. The recombinant DNA molecules are then put into a bacteria strain (usually *E. coli*) which produces several identical copies by transformation. Transformation is the DNA uptake mechanism possessed by bacteria. However, only one recombinant DNA molecule can be cloned within a single bacteria cell, so each clone is of just one DNA insert.

Separation and Detection

In separation and detection DNA and mRNA are isolated from cells (the separation) and then detected simply by the isolation. Cell cultures are also grown to provide a constant supply of cells ready for isolation.

Cell Cultures

A cell culture for molecular genetics is a culture that is grown in artificial conditions. Some cell types grow well in cultures such a skin cells, but other cells are not as productive in cultures. There are different techniques for each type of cell, some only recently being found to foster growth in stem and nerve cells. Cultures for molecular genetics are frozen in order to preserve all copies of the gene specimen and thawed only when needed. This allows for a steady supply of cells.

DNA Isolation

DNA isolation extracts DNA from a cell in a pure form. First, the DNA is separated from cellular components such as proteins, RNA, and lipids. This is done by placing the chosen cells in a tube with a solution that mechanically, chemically, breaks the cells open. This solution contains enzymes, chemicals, and salts that breaks down the cells except for the DNA. It contains enzymes to dissolve proteins, chemicals to destroy all RNA present, and salts to help pull DNA out of the solution.

Next, the DNA is separated from the solution by being spun in a centrifuge, which allows the DNA to collect in the bottom of the tube. After this cycle in the centrifuge the solution is poured off and the DNA is resuspended in a second solution that makes the DNA easy to work with in the future.

This results in a concentrated DNA sample that contains thousands of copies of each gene. For large scale projects such as sequencing the human genome, all this work is done by robots.

mRNA Isolation

Expressed DNA that codes for the synthesis of a protein is the final goal for scientists and this expressed DNA is obtained by isolation mRNA (Messenger RNA). First, laboratories use a normal cellular modification of mRNA that adds up to 200 adenine nucleotides to the end of the molecule (poly(A) tail). Once this has been added, the cell is ruptured and its cell contents are exposed to synthetic beads that are coated with thymine string nucleotides. Because Adenine and Thymine pair together in DNA, the poly(A) tail and synthetic beads are attracted to one another, and once they bind in this process the cell components can be washed away without removing the mRNA. Once the mRNA has been isolated, reverse transcriptase is employed to convert it to single-stranded DNA, from which a stable double-stranded DNA is produced using DNA polymerase. Complementary DNA (cDNA) is much more stable than mRNA and so, once the double-stranded DNA has been produced it represents the expressed DNA sequence scientists look for.

The Molecular Genetics Project

The Human Genome Project is a molecular genetics project that began in 1990 and was projected to take fifteen years to complete. However, because of technological advances the progress of the project was advanced and the project finished in 2003, taking only thirteen years. The project was started by the U.S. Department of Energy and the National Institutes of Health in an effort to reach six set goals. These goals included:

1. identifying 20,000 to 25,000 genes in human DNA (although initial estimate were approximately 100,000 genes),
2. determining sequences of chemical based pairs in human DNA,
3. storing all found information into databases,
4. improving the tools used for data analysis,
5. transferring technologies to private sectors, and
6. addressing the ethical, legal, and social issues (ELSI) that may arise from the projects.

The project was worked on by eighteen different countries including the United States, Japan, France, Germany, and the United Kingdom. The collaborative effort resulted in the discovery of the many benefits of molecular genetics. Discoveries such as molecular medicine, new energy sources and environmental applications, DNA forensics, and livestock breeding, are only a few of the benefits that molecular genetics can provide.

POPULATION GENETICS

Population genetics is the study of the allele frequency distribution and change under the influence of the four evolutionary forces: natural selection, genetic drift, mutation and gene flow. It also takes account of population subdivision and population structure in space. As such, it attempts to explain such phenomena as adaptation and speciation. Population genetics was a vital ingredient in the modern evolutionary synthesis, its primary founders were Sewall Wright, J. B. S. Haldane and R. A. Fisher, who also laid the foundations for the related discipline of quantitative genetics.

Scope and Theoretical Considerations

Perhaps the most significant "formal" achievement of the modern evolutionary synthesis has been the framework of mathematical population genetics. Indeed some authors (Beatty 1986) would argue that it does *define* the core of the modern synthesis.

Lewontin (1974) outlined the theoretical task for population genetics. He imagined two spaces: a "genotypic space" and a "phenotypic space". The challenge of a *complete* theory of population genetics is to provide a set of laws that predictably map a population of genotypes (G_1) to a phenotype space (P_1), where selection takes place, and another set of laws that map the resulting population (P_2) back to genotype space (G_2) where Mendelian genetics can predict the next generation of genotypes, thus completing the cycle. Even leaving aside for the moment the non-Mendelian aspects revealed by molecular genetics, this is clearly a gargantuan task. Visualizing this transformation schematically:

$$G_1 \rightarrow^{T_2} P_1 \rightarrow^{T_2} P_2 \rightarrow^{T_3} G_2 \rightarrow^{T_4} G_2 \rightarrow \ldots.$$

(adapted from Lewontin 1974, p. 12).

T^1 represents the genetic and epigenetic laws, the aspects of functional biology, or development, that transform a genotype into phenotype. We will refer to this as the "genotype-phenotype map". T^2 is the transformation due to natural selection, T^3

are epigenetic relations that predict genotypes based on the selected phenotypes and finally T^4 the rules of Mendelian genetics.

In practice, there are two bodies of evolutionary theory that exist in parallel, traditional population genetics operating in the genotype space and the biometric theory used in plant and animal breeding, operating in phenotype space. The missing part is the mapping between the genotype and phenotype space. This leads to a "sleight of hand" (as Lewontin terms it) whereby variables in the equations of one domain, are considered parameters or *constants*, where, in a full-treatment they would be transformed themselves by the evolutionary process and are in reality *functions* of the state variables in the other domain. The "sleight of hand" is assuming that we know this mapping. Proceeding as if we do understand it is enough to analyze many cases of interest. For example, if the phenotype is almost one-to-one with genotype (sickle-cell disease) or the time-scale is sufficiently short, the "constants" can be treated as such; however, there are many situations where it is inaccurate.

Population Geneticists

The three founders of population genetics were the Britons R.A. Fisher and J.B.S. Haldane and the American Sewall Wright. Fisher and Wright had some fundamental disagreements and a controversy about the relative roles of selection and drift continued for much of the century between the Americans and the British. The Frenchman Gustave Malécot was also important early in the development of the discipline. John Maynard Smith was Haldane's pupil, whilst W.D. Hamilton was heavily influenced by the writings of Fisher. The American George R. Price worked with both Hamilton and Maynard Smith. On the American side, Richard Lewontin and the Japanese Motoo Kimura were heavily influenced by Wright.

QUANTITATIVE GENETICS

Quantitative genetics is the study of continuous traits (such as height or weight) and its underlying mechanisms. It is effectively an extension of simple Mendelian inheritance in that the combined effect of the many underlying genes results in a continuous distribution of phenotypic values.

History

The field was founded, in evolutionary terms, by the originators of the modern synthesis, R.A. Fisher, Sewall Wright and J.B.S. Haldane, and aimed to predict the response to selection given data on the phenotype and relationships of individuals.

Analysis of Quantitative trait loci, or QTL, is a more recent addition to the study of quantitative genetics. A QTL is a region in the genome that affects the trait or traits of interest. Quantitative trait loci approaches require accurate phenotypic, pedigree and genotypic data from a large number of individuals.

Traits

Quantitative genetics is not limited to continuous traits, but to all traits that are determined by many genes. This includes:

- Continuous traits are quantitative traits with a continuous phenotypic range. They are often polygenic, and may also be influenced significantly by environmental effects.
- Meristic traits or other ordinal numbers are expressed in whole numbers, such as number of offspring, or number of bristles on a fruit fly. These traits can be either treated as approximately continuous traits or as threshold traits.
- Some qualitative traits can be treated as if they have an underlying quantitative basis, expressed as a threshold trait (or multiple thresholds). Some human diseases (such as, schizophrenia) have been studied in this manner.

Basic Principles

The phenotypic value (P) of an individual is the combined effect of the genotypic value (G) and the environmental deviation (E):

$$P = G + E$$

The genotypic value is the combined effect of all the genetic effects, including nuclear genes, mitochondrial genes and interactions between the genes. It is therefore often subdivided in an additive (A) and a dominance component (D). The additive effect described the cumulative effect of the individual genes, while the dominance effect is the result of interactions between those genes. The environmental deviation can be subdivided in a pure environmental component (E) and an interaction factor (I) describing the interaction between genes and the environment. This can be described as:

$$P = A + D + E + I$$

The contribution of those components cannot be determined in a single individual, but they can be estimated for whole populations by estimating the variances for those components, denoted as:

$$VP = VA + VD + VE + VI$$

The heritability of a trait is the proportion of the total (i.e. phenotypic) variation (V_P) that is explained by the genetic variation. This is the total genetic variation (V_G) in broad sense heritabilities (H^2), while only the additive genetic variation (V_A) is used for narrow sense heritabilities (h^2), often simply called heritability. The latter gives an indication how a trait will respond to natural or artificial selection.

Resemblance Between Relatives

Central in estimating the variances for the various components is the principle of relatedness. A child has a father and a mother. Consequently, the child and father share 50% of their alleles, as do the child and the mother. However, the mother and father normally do not share genes as a result of shared ancestors. Similarly, two full siblings share also on average 50% of the alleles with each other, while half sibs share only 25% of their alleles. This variation in relatedness can be used to estimate which proportion of the total phenotypic variance (VP) is explained by the above-mentioned components.

$$\textit{Genetic correlation} = \frac{Cov(A_1, A_2)}{\sqrt{V_{A_1} * V_{A_2}}}$$

Correlated Traits

Although some genes have only an effect on a single trait, many genes have an effect on various traits. Because of this, a change in a single gene will have an effect on all those traits. This is calculated using covariances, and the phenotypic covariance (CovP) between two traits can be partitioned in the same way as the variances described above. The genetic correlation is calculated by dividing the covariance between the additive genetic effects of two traits by the square root of the product of the variances for the additive genetic effects of the two traits:

REFERENCES

Cain A.J. and W.B. Provine (1992). Genes and ecology in history. In: R.J. Berry, T.J. Crawford and G.M. Hewitt (eds). *Genes in Ecology*. Blackwell Scientific: Oxford. (Provides a good historical background)

Conner, J.K. and Hartl, D. L. "A Primer of Ecological Genetics". Sinauer Associates, Inc.; Sunderland, Mass. (2004) Provides basic and intermediate level processes and methods.

J. Beatty. "The synthesis and the synthetic theory" in Integrating Scientific Disciplines, edited by W. Bechtel and Nijhoff. Dordrecht, 1986.

Luigi Luca Cavalli-Sforza. Genes, Peoples, and Languages. North Point Press, 2000.

Luigi Luca Cavalli-Sforza et al. The History and Geography of Human Genes. Princeton University Press, 1994.

James F. Crow and Motoo Kimura. Introduction to Population Genetics Theory. Harper & Row, 1972.

Warren J Ewens. Mathematical Population Genetics. Springer-Verlag New York, Inc., 2004.

John H. Gillespie Population Genetics: A Concise Guide, Johns Hopkins Press, 1998.

Richard Halliburton. Introduction to Population Genetics. Prentice Hall, 2004

Daniel Hartl. Primer of Population Genetics, 3rd edition. Sinauer, 2000.

Daniel Hartl and Andrew Clark. Principles of Population Genetics, 3rd edition. Sinauer, 1997.

Richard C. Lewontin. The Genetic Basis of Evolutionary Change. Columbia University Press, 1974.

William B. Provine. The Origins of Theoretical Population Genetics. University of Chicago Press. 1971.

Spencer Wells. The Journey of Man. Random House, 2002.

Spencer Wells. Deep Ancestry: Inside the Genographic Project. National Geographic Society, 2006.

Falconer, D.S. & Mackay TFC (1996). Introduction to Quantitative Genetics. Fourth edition. Addison Wesley Longman, Harlow, Essex, UK.

Roff DA (1997). Evolutionary Quantitative Genetics. Chapman & Hall, New York.

Lynch M & Walsh B (1998). Genetics and Analysis of Quantitative Traits. Sinauer, Sunderland, MA.

4

Gene Related Processes, Genetic Changes and Cloning

GENE POOL

In population genetics, a gene pool is the complete set of unique alleles in a species or population. A large gene pool indicates extensive genetic diversity, which is associated with robust populations that can survive bouts of intense selection. Meanwhile, low genetic diversity (see inbreeding and population bottlenecks) can cause reduced biological fitness and an increased chance of extinction.

When all individuals in a population are identical with regard to a particular phenotypic trait they are known as monomorphic. When the individuals show several variants of a particular trait they are said to be polymorphic .

GENE FLOW

In population genetics, gene flow (also known as gene migration) is the transfer of alleles of genes from one population to another.

Migration into or out of a population may be responsible for a marked change in allele frequencies (the proportion of members carrying a particular variant of a gene). Immigration may also result in the addition of new genetic variants to the established gene pool of a particular species or population.

There are a number of factors that affect the rate of gene flow between different populations. One of the most significant factors is mobility, as greater mobility of an individual tends to give it greater migratory potential. Animals tend to be more mobile than plants, although pollen and seeds may be carried great distances by animals or wind.

Maintained gene flow between two populations can also lead to a combination of the two gene pools, reducing the genetic variation between the two groups. It is for this reason that gene flow strongly acts against speciation, by recombining the gene pools of the groups, and thus, repairing the developing differences in genetic variation that would have led to full speciation and creation of daughter species.

Barrier to Gene Flow

Physical barriers to gene flow are usually, but not always, natural. They may include impassable mountain ranges, oceans, or vast deserts. In some cases, they can be artificial, man-made barriers, such as the Great Wall of China, which has hindered the gene flow of native plant populations. Samples of the same species which grow on either side have been shown to have developed genetic differences, because there is no gene flow to provide recombination of the gene pools.

Barriers to gene flow need not always to be physical. Species can live in the same environment, yet show very limited gene flow due to limited hybridization or hybridization yielding unfit hybrids.

Gene Flow in Humans

Gene flow has been observed in humans. For example, in the United States, gene flow was observed between a white European population and a black West African population, which were recently brought together. In West Africa, where malaria is prevalent, the Duffy antigen provides some resistance to the disease, and this allele is thus present in nearly all of the West African population. In contrast, Europeans have either the allele Fya or Fyb, because malaria is almost nonexistent. By measuring the frequencies of the West African and European groups, scientists found that the allele frequencies became mixed in each population because of movement of individuals. It was also found that this gene flow between European and West African groups is much greater in the Northern U.S. than in the South.

Gene Flow between Species

Genes flow can occur between species, either through hybridization or gene transfer from bacteria or virus to new hosts.

Gene transfer, defined as the movement of genetic material across species boundaries, which includes horizontal gene transfer, antigenic shift, and reassortment is sometimes an important source of genetic variation. Viruses can transfer genes between species. Bacteria can incorporate genes from other dead bacteria, exchange genes with living bacteria, and can exchange plasmids across species boundaries. "Sequence comparisons suggest recent horizontal transfer of many genes among diverse species including across the boundaries of phylogenetic "domains". Thus determining the phylogenetic history of a species can not be done conclusively by determining evolutionary trees for single genes."

Biologist Gogarten suggests "the original metaphor of a tree no longer fits the data from recent genome research". Biologists [should] instead use the metaphor of a mosaic to describe the different histories combined in individual genomes and use the metaphor of an intertwinded net to visualize the rich exchange and cooperative effects of horizontal gen transfer.

"Using single genes as phylogenetic markers, it is difficult to trace organismal phylogeny in the presence of HGT [horizontal gene transfer]. Combining the simple coalescence model of cladogenesis with rare HGT [horizontal gene transfer] events suggest there was no single last common ancestor that contained all of the genes ancestral to those shared among the three domains of life. Each contemporary molecule has its own history and traces back to an individual molecule cenancestor. However, these molecular ancestors were likely to be present in different organisms at different times."

Genetic Pollution

Purebred naturally evolved region specific wild species can be threatened with extinction in a big way through the process of Genetic Pollution i.e. uncontrolled hybridization, introgression and Genetic swamping which leads to homogenization or replacement of local genotypes as a result of either a numerical and/or fitness advantage of introduced plant or animal. Nonnative species can bring about a form of extinction of native plants and animals by hybridization and introgression either through purposeful introduction by humans or through habitat modification, bringing previously isolated species into contact. These phenomena can be especially detrimental for rare species coming into contact with more abundant ones where the abundant ones can interbreed with them swamping the entire rarer gene pool creating hybrids thus driving the entire original purebred native stock to complete extinction. Attention has to be focused on the extent of this under appreciated problem that is not always apparent from morphological (outward appearance) observations alone. Some degree of gene flow may be a normal, evolutionarily constructive process, and all constellations of genes and genotypes cannot be preserved however, hybridization with or without introgression may, nevertheless, threaten a rare species' existence.

Models of Gene Flow

Models of gene flow can be derived from population genetics, e.g. Sewall Wright's neighborhood model, Wright's island model and the stepping stone model.

Gene Flow Mitigation

When cultivating genetically modified (GM) plants or livestock, it becomes necessary to prevent "genetic pollution" i.e. their genetic modification from reaching other conventionally hybridized or wild native plant and animal populations by using gene flow mitigation usually through unintentional cross pollination and crossbreeding. Reasons to limit gene flow may include biosafety or agricultural co-existence, in which GM and non-GM cropping systems work side by side.

Scientists in several large research programmes are investigating methods of limiting gene flow in plants. Among these programmes are Transcontainer, which investigates methods for biocontainment, SIGMEA, which focuses on the biosafety of genetically modified plants, and Co-Extra, which studies the co-existence of GM and non-GM product chains.

Generally, there are three approaches to gene flow mitigation: keeping the genetic modification out of the pollen, preventing the formation of pollen, and keeping the pollen inside the flower.

- The first approach requires transplastomic plants. In transplastomic plants, the modified DNA is not situated in the cell's nucleus but is present in plastids, which are cellular compartments outside the nucleus. An example for plastids are chloroplasts, in which photosynthesis occurs. In some plants, the pollen does not contain plastids and, consequently, any modification located in plastids cannot be transmitted by the pollen.
- The second approach relies on male sterile plants. Male sterile plants are unable to produce functioning flowers and therefore cannot release viable pollen. Cytoplasmic male sterile plants are known to produce higher yields. Therefore, researchers are trying to introduce this trait to genetically modified crops.
- The third approach works by preventing the flowers from opening. This trait is called cleistogamy and occurs naturally in some plants. Cleistogamous plants produce flowers which either open only partly or not at all. However, it remains unclear how reliable cleistogamy is for gene flow mitigation: a Co-Extra research project on rapeseed investigating the matter has published preliminary results which cast doubt on the attainment of a high degree of reliability.

GENETIC DRIFT

In population genetics, genetic drift (or more precisely allelic drift) is the statistical effect that results from the influence that chance has on the survival of alleles (variants of a gene). The effect may cause an allele, and the biological traits that it confers, to become more common or rare over successive generations. Whereas natural selection defines the tendency of alleles to become more common or less common in a population over time as a result of the alleles' effects on adaptive and reproductive success, genetic drift is the fundamental tendency of an allele distribution in a population to statistically alter over time due to random events. Genetic drift is one of the primary mechanisms of biological evolution.

Chance affects the commonality or rarity of an allele because an organism's survival and reproductive success is subject to other factors besides adaptive pressures. When chance events preserve the survival of randomly selected organisms of a given population, and the resulting allele frequency of the surviving group differs statistically from allele frequencies in that original population, the evolution occurs as a statistical phenomenon rather than selective phenomenon. When in such cases the frequency distribution in the survivors differs from the frequency distribution of the original population, the evolutionary result is due to the effect of sampling error and probability.

Genetic drift depends strongly on population size as a consequence of the law of large numbers. When many individuals carry a particular allele, and all face equal odds, the number of offspring they collectively produce will only slightly differ from the expected value, which is the expected average per individual times the number of individuals. But with a small effective breeding pool, a departure from the norm in even one individual can cause a disproportionately greater deviation from the expected result. Therefore small populations are more subject to genetic drift than large ones. This is also the basis for the founder effect, a proposed mechanism of speciation.

By definition, genetic drift has no preferred direction. A neutral allele may be expected to increase or decrease in any given generation with equal probability. Given sufficiently long time, however, the mathematics of genetic drift (cf. Galton-Watson process) predict the allele will either die out or be present in 100% of the population, after which time there is no random variation in the associated gene. Thus genetic drift tends to sweep gene variants out of a population over time, such that all members of a species would eventually be homozygous for this gene. In this regard, genetic drift opposes genetic mutation which introduces novel variants into the population according to its own random processes.

Allele Frequencies

From the perspective of population genetics, drift is a "sampling effect." To illustrate: on average, coins turn up heads or tails with equal probability. Yet just a few tosses in a row are unlikely to produce heads and tails in equal number. The numbers are no more likely to be exactly equal for many tosses in a row, but the discrepancy in number can be very small (in percentage terms). As an example, ten tosses turn up at least 70% heads about once in every six tries, but the chance of a hundred tosses in a row producing at least 70% heads is only about one in 25,000.

Similarly, in a breeding population, if an allele has a frequency of *p*, probability theory dictates that (if natural selection is not acting) in the following generation, a fraction *p* of the population will inherit that particular allele. However, as with the coin toss above, allele frequencies in real populations are not probability distributions; rather, they are a random sample, and are thus subject to the same statistical fluctuations (sampling error).

When the alleles of a gene do not differ with regard to fitness, on average the number of carriers in one generation is proportional to the number of carriers in the previous generation. But the average is never tallied, because each generation parents the next one only once. Therefore the frequency of an allele among the offspring often differs from its frequency in the parent generation. In the offspring generation, the allele might therefore have a frequency *p'*, slightly different from *p*. In this situation, the allele frequencies are said to have drifted. Note that the frequency of the allele in subsequent generations will now be determined by the new frequency *p'*, meaning that drift is a memoryless process and may be modeled as a Markov process.

As in the coin toss example above, the size of the breeding population (the effective population size) governs the strength of the drift effect. When the effective population size is small, genetic drift will be stronger.

Drifting alleles usually have a finite lifetime. As the frequency of an allele drifts up and down over successive generations, eventually it drifts until fixation - that is, it either reaches a frequency of zero, and disappears from the population, or it reaches a frequency of 100% and becomes the only allele in the population. Subsequent to the latter event, the allele frequency can only change by the introduction of a new allele by a new mutation.

The lifetime of an allele is governed by the effective population size. In a very small population, only a few generations might be required for genetic drift to result in fixation. In a large population, it would take many more generations. On average, an allele will be fixed in $4Ne$ generations, where Ne is the effective population size.

According to the Hardy-Weinberg Principle, which holds that allele frequencies in a gene pool will not change over time, a population must be sufficiently large to prevent genetic drift from changing allele frequencies over time. This is why the law is unstable in a small population.

Drift Versus Selection

Genetic drift and natural selection rarely occur in isolation from each other; both forces are always at play in a population. However, the degree to which alleles are affected by drift and selection varies according to circumstance.

In a large population, where genetic drift occurs very slowly, even weak selection on an allele will push its frequency upwards or downwards (depending on whether the allele is beneficial or harmful). However, if the population is very small, drift will predominate. In this case, weak selective effects may not be seen at all as the small changes in frequency they would produce are overshadowed by drift.

Genetic Drift in Populations

Drift can have profound and often bizarre effects on the evolutionary history of a population. These effects may be at odds with the survival of the population.

In a population bottleneck, where the population suddenly contracts to a small size, genetic drift can result in sudden and dramatic changes in allele frequency that occur independently of selection. According to the Toba catastrophe theory this occurred in the history of human evolution. In such instances, many beneficial adaptations may be eliminated even if population later grows large again.

Similarly, migrating populations may see a founder effect, where a few individuals in the originating generation with a rare allele can produce a population that has allele frequencies that seem at odds with natural selection. Founder's effects are sometimes held to be responsible for high frequencies of some genetic diseases.

GENETIC POLLUTION

Genetic pollution, genetic contamination or genetic swamping happens when an original set of naturally evolved (wild) region specific genes or gene pool of a wild animal or plant variety on its own becomes hybridized with a domesticated / feral one, or with the genes of another wild introduced / nonnative / invasive variety.

While intentional controlled crossbreeding between two genetically distinct varieties can be described as hybridization, genetic pollution can be described as unintentional uncontrolled crossbreeding and gene flow between any such two whether they be two wild or domestic / feral genetically distinct varieties or a wild and a domestic / feral variety, where wild includes introduced / nonnative / invasive varieties and domestic / feral includes artificially genetically engineered (GE) and genetically modified (GM) varieties.

There is special concern today about the increased occurrence of genes from artificially genetically engineered (GE) and genetically modified (GM) varieties finding their way into other domesticated varieties derived through relatively natural process of controlled selective breeding, and into their native landraces and wild occurring relatives through the process of genetic pollution when let loose in the natural environment. It is also being warned that these artificially engineered and modified genes will eventually find their way even into unrelated varieties in nature soon.

Usually genetic pollution i.e. uncontrolled hybridization, introgression and genetic swamping happens as a result and a consequence of mankind's interaction with natural environment and a lack of foresight but in rare instances it has also been observed happening naturally more commonly in case of closely related subspecies of wild plants whose ranges overlap forming hybrid zones making it easier for insects to cross pollinate them.

In most cases nature has its own interspecies genetic barriers to guard against genetic pollution to keep species distinct. When rarely hybridization does occur naturally as in hybrid zones where the ranges of closely related wild species overlap, the hybrid crosses produced, even though they may display hybrid vigour (heterosis) in the first generation (F1 hybrid), are in the long run less fit than the two parent species which have evolved over hundreds of thousands of years specializing in exploiting their own particular niche in nature. It is extremely rare that the hybrids ever become fitter than the two wild parent species so that natural selection may then favor these individuals and it is even rarer that reproductive isolation is ever achieved to lead to the birth of a new species through the process known as hybrid speciation.

"Genetic Pollution" - A Term that has Gained Popular Use and Increased Awareness

The term genetic pollution was popularized by environmentalist Jeremy Rifkin in his 1998 book The Biotech Century, he has used it for explaining the risk of genes from genetically modified organisms (GMOs) getting dispersed into the natural environment as a result of them breeding with wild relatives, or even with unrelated species. The term has gained popular use within the environmental movement and is increasingly turning up in newspaper reports. Though commonly found used in scientific reports worldwide today and within the scientific community; pro biotechnology, pro GE and pro GM scientists, companies and developed countries, though obsoletely yet often still think of "genetic pollution" as a non scientific term used to provoke an emotional response, and gene flow as a more scientific term to describe the natural or human caused movement of genes between living organisms. However the term is widely used by scientists today within the Food, Agriculture,

Horticulture and Animal husbandry sector and outside in general Environmental, Biodiversity and Wildlife conservation.

Food and Agriculture Organization of the United Nations (FAO) Glossary describes genetic pollution as "Uncontrolled spread of genetic information (frequently referring to transgenes) into the genomes of organisms in which such genes are not present in nature."

Canadian Broadcasting Corporation's CBC News Online in it's article "INDEPTH: LANGUAGE - 5,000 new words" dated July 26, 2004 mentions Canada has 5,000 new official words that have been added to the recently released Second Edition of the Canadian Oxford dictionary. In the random sampling of some of the terms provided "genetic pollution" is described as "the spread of altered genes from genetically engineered organisms to other, non-engineered organisms, esp. by cross-pollination".

TRAFFIC is the international wildlife trade monitoring network which works to ensure that trade in wild plants and animals is not a threat to the conservation of nature. It is a joint programme of WWF and IUCN - The World Conservation Union. Spreading awareness about the harmful effects of introduced non native invasive species it explains "Wildlife trade can also cause indirect harm through: Introducing invasive species which then prey on, or compete with, native species. Invasive species are those which have been brought into a wild area where they do not occur naturally and which have a negative affect on native species. Invasive species have been a major cause of extinction throughout the world in the past few hundred years. Some of them prey on native wildlife, compete with it for resources, or spread disease, while others may hybridize with native species, causing "genetic pollution". In these ways, invasive species are as big a threat to the balance of nature as the direct overexploitation by humans of some species. Wildlife traders have purposely introduced many invasive species; examples include the American Mink, the Red-eared Terrapin and countless plant species."

Joint Nature Conservation Committee (JNCC) is the statutory adviser to Government of United Kingdom and international nature conservation. Its work contributes to maintaining and enriching biological diversity, conserving geological features and sustaining natural systems. Educating about the harmful effects of the introduction of invasive/non-native species it advises "Occasionally non-native species can reproduce with native species and produce hybrids, which will alter the genetic pool (a process called genetic pollution), which is an irreversible change."

Endangers Wild Species and Quickens Their Extinction

Pure-bred naturally evolved region specific wild species can be threatened with extinction in a big way through the process of genetic pollution i.e. uncontrolled hybridization, introgression and genetic swamping which leads to homogenization or replacement of local genotypes as a result of either a numerical and/or fitness advantage of introduced plant or animal. Non-native species can bring about a form of extinction of native plants and animals by hybridization and introgression either through purposeful introduction by humans or through habitat modification, bringing previously isolated species into contact. These phenomena can be especially detrimental for rare species coming into contact with more abundant ones where the abundant ones can interbreed with them swamping the entire rarer gene pool creating hybrids thus driving the entire original pure-bred native stock to complete extinction. Attention has to be focused on the extent of this under appreciated problem that is not always apparent from morphological (outward appearance) observations alone. Some degree of gene flow may be a normal, evolutionarily constructive process, and all constellations of genes and genotypes cannot be preserved however, hybridization with or without introgression may, nevertheless, threaten a rare species' existence.

Widespread genetic pollution also leads to weakening of the naturally evolved (wild) region specific gene pool leading to weaker hybrid animals and plants which are not able to cope with natural environs over the long run and fast tracks them towards final extinction.

The gene pool of a species or a population is the complete set of unique alleles that would be found by inspecting the genetic material of every living member of that species or population. A large gene pool indicates extensive genetic diversity, which is associated with robust populations that can survive bouts of intense selection. Meanwhile, low genetic diversity (see inbreeding and population bottlenecks) can cause reduced biological fitness and an increased chance of extinction amongst the reducing population of pure-bred individuals from a species.

Causes

Nowadays, with several domesticated and hybridized animals and plants living in towns and villages in and all around forests which still hold naturally evolved region specific original wild ancestor species and subspecies of their kind, genetic pollution has become a major threat to naturally evolved wild populations and their natural genetic make-up. Wild populations of animals and plants have evolved naturally over millions of years through a process of natural selection as apposed to human controlled selective breeding or artificial selection in animal husbandry and agriculture for desirable traits from the human point of view. These naturally evolved wild species (wildlife) and region specific subspecies are most adaptable to living in the particular regions of earth where they exist presently. They have scientifically the most endurable and viable genetic make-up to live where they live. Genetic pollution leads to hybridized animals and plants which, even though they may look exactly the same, are not as genetically strong as their naturally evolved region specific wild ancestors who can survive under natural wild conditions indefinitely without any human help or husbandry and have high adaptability to local climatic conditions and a high immunity to local natural diseases.

Throughout history with the movement and immigration of people from one part of the world to another, intentionally or unintentionally non-native species have spread creating havoc on indigenous wild animals and plants and local ecosystems. Non-native species were introduced intentionally into new regions because of aesthetic reasons or for hunting & sport and sometimes for emotional reasons as people missed them in their adopted homelands where they had settled. These non-native species and subspecies have become another common source of genetic pollution as they readily cross-breed with closely related often much rarer indigenous forms.

In modern times even the national and international zoo associations have realized how important it is to maintain pure-bred naturally evolved region specific subspecies of wild animals which may one day have a chance to be reintroduced back and survive in the wild in their original home range. It has now become mandatory for zoos to maintain detailed breeding records of their wild animal stock, keep detailed studbooks and have registered breeding programs. Random hybridization of wild animals with subspecies from different regions was a common practice which plagued all of world's zoos until recently.

Usually genetic pollution happens as a result and a consequence of mankind's activities and interaction with natural environment and a lack of foresight but in rare instances it has also been observed happening naturally more commonly in case of closely related subspecies of plants whose ranges overlap forming hybrid zones making it easier for insects to cross pollinate them.

Significance to Humans

Amongst other threats it is also important to give due importance to and guard against genetic pollution so that we can conserve the naturally evolved region specific wild gene stock and genetic make-up of wild animals and wild plants for maintaining the health of natural ecosystems and the environment in general. Scientists now, also consider this naturally evolved biodiversity to be a valuable source of strong genes which in the future may be used on a continual basis to hybridize domesticated varieties, to make them even stronger and more resistant to climate and diseases, thus leading to an ongoing improvement in our food security and medicines etc. for us humans ourselves.

Examples

Native captive Asiatic Lions in Indian zoos until recently were genetically polluted with genes of Circus confiscated African Lions which had been randomly hybridized with them leading to widespread genetic pollution in the captive Asiatic Lion population. Once discovered this led to the complete shut down of the European (EEP) and the American endangered species registered breeding programs (SSP) for Asiatic Lions as the founder animals originally imported from India were ascertained to be genetically polluted with the genes of African lions. Since then India has corrected its mistake and now breeds only pure native Asiatic Lions which are critically endangered with extinction and has helped revive the European endangered species registered breeding program (EEP) for Asiatic Lions but the American SSP which completely shut down in early 80's has yet to be revived and receive pure bred Asiatic Lions from India to form a new founder population for breeding in the zoos on the American continent.

Wild Jungle fowl, specifically Red Junglefowl in India and South Asia are the ancestor of all Poultry Chickens. These are thought to be facing a serious threat of extinction because of genetic pollution which is occurring at the edge of forests where domesticated free ranging chickens are commonly kept in bordering villages and towns, these meet and breed with their wild cousins indiscriminately genetically polluting their ancestors.

The Wild Asian Water Buffalo is facing a serious threat of extinction because of genetic pollution when it comes into contact with common abundant Domestic Asian Water Buffalo, which live in villages and towns all around forests. The domesticated animals daily come to graze within forests which have been designated as Wildlife Sanctuaries and National Parks for their very rare wild ancestors and interbreed with them genetically swamping the wild gene pool raising fears that no pure Wild Asian Water Buffalo may be alive today.

Many herds of surviving American Bison, even though they look exactly the same, were hybridized with genes of beef cattle by landowners in a short-sighted effort to improve their meat quality (see Cattalo), leading to widespread genetic pollution in the surviving bisons in America. The current American Bison population has been growing rapidly and is estimated at 350,000, but this is compared to an estimated 60–100 million in the mid-19th century. Most current herds, however are genetically polluted or partly crossbred with cattle hence are in fact what are called "beefalo;" today there are only four genetically unmixed herds and only one that is also free of brucellosis: it roams Wind Cave National Park. A founder population from the Wind Cave herd was recently established in Montana by the World Wildlife Fund.

Tara, a hand-reared supposedly Bengal tigress acquired from Twycross Zoo in England in July 1976 was trained by Billy Arjan Singh and released to the wild in Dudhwa National Park, India with the permission of India's then Prime Minister

Indira Gandhi in an attempt to prove the experts wrong that zoo bred hand reared Tigers can ever be released in the wild with success. In the 1990s, some tigers from Dhudhwa were observed which had the typical appearance of Siberian tigers: white complexion, pale fur, large head and wide stripes. With recent advances in science it was subsequently found that Siberian Tigers genes have polluted the otherwise pure Bengal Tiger gene pool of Dudhwa National Park. It was proved later that Twycross Zoo had been irresponsible and maintained no breeding records and had given India a hybrid Siberian-Bengal Tigress instead. Dudhwa tigers constitute about 1% of India's total wild population, but the possibility exists of this genetic pollution spreading to other tiger groups, at its worst, this could jeopardize the Bengal tiger as a distinct subspecies.

Release of feral Mallard Ducks worldwide is creating havoc on indigenous waterfowl, unlike wild Mallards these don't migrate and stay back in the local breeding season and interbreed with indigenous rare wild ducks devastating local populations of closely related species through genetic pollution by producing fertile offspring. Complete hybridization of various species of rare wild duck gene pools could result in the extinction of many indigenous waterfowl. Wild Mallard itself is the ancestor of most domestic ducks and their naturally evolved wild gene pool gets genetically polluted in turn by the domestic and feral populations.

Conventional Hybridization a Threat to Biodiversity and Food Security

In agriculture and animal husbandry, green revolution popularized the use of conventional hybridization to increase yield many folds by creating "high-yielding varieties". Often the handful of breeds of plants and animals hybridized originated in developed countries and were further hybridized with local varieties, in the rest of the developing world, to create high yield strains resistant to local climate and diseases. Local governments and industry since have been pushing hybridization with such zeal that several of the wild and indigenous breeds evolved locally over thousands of years having high resistance to local extremes in climate and immunity to diseases etc. have already become extinct or are in grave danger of becoming so in the near future. Due to complete disuse because of un-profitability and uncontrolled intentional, compounded with unintentional cross pollination and cross-breeding (genetic pollution) formerly huge gene pools of various wild and indigenous breeds have collapsed causing widespread genetic erosion and genetic pollution resulting in great loss in genetic diversity and biodiversity as a whole.

A Genetically Modified Organism (GMO) is an organism whose genetic material has been altered using the genetic engineering techniques generally known as recombinant DNA technology. Genetic engineering today has become another serious and alarming cause of genetic pollution because artificially created and genetically engineered plants and animals in laboratories, which could never have evolved in nature even with conventional hybridization, can live and breed on their own and what is even more alarming interbreed with naturally evolved wild varieties. Terminator technology represents one currently neglected technology that could prevent the spread of genetic material from GMOs. Genetically Modified (GM) crops today have become a common source for genetic pollution, not only of wild varieties but also of other domesticated varieties derived from relatively natural hybridization.

It is being said that genetic erosion coupled with genetic pollution is destroying that needed unique genetic base thereby creating an unforeseen hidden crisis which will result in a severe threat to our food security for the future when diverse genetic material will cease to exist to be able to further improve or hybridize weakening food crops and livestock against more resistant diseases and climatic changes.

Natural Barriers to Genetic Pollution

It is common knowledge that plants cannot reproduce with animals, nor can fishes with reptiles or birds with insects and so on. Different species within the same animal or plant kingdoms also can not reproduce with each other (by definition) and are not attracted to each other in the wild for the purpose of reproduction. Species which fall under the same broad families like donkeys and horses or tigers and lions will mate with each other only in captivity and only if raised together from young ones but most interspecies hybrids are born without functioning reproductive systems and hence can not reproduce any further. Common example of the most famous interspecies hybrid created and utilized by man is the cross between donkeys and horses, the mule which is commonly utilized as a hardy draught animal in rough terrain but mules can not reproduce any further. The term “mule” (Latin mulus) was formerly applied to the offspring of any two creatures of different species - in modern usage, a “hybrid”. Mules are almost always sterile (see fertile mules for rare cases).

Interspecific hybrids like the mule are bred by mating two species, normally from within the same genus. The offspring display traits and characteristics of both parents. The offspring of an interspecific cross very often are sterile, this hybrid sterility prevents the movement of genes from one species to the other, keeping both species distinct. Sterility is often attributed to the different number of chromosomes the two species have, for example donkeys have 62 chromosomes, while horses have 64 chromosomes, and mules and hinnies, the offspring thus have 63 chromosomes which cannot evenly divide. Mules, hinnies, and other normally sterile interspecific hybrids cannot produce viable gametes because the extra chromosome cannot make a homologous pair at meiosis, meiosis is disrupted, and viable sperm and eggs are not formed.

Most often other mechanisms are used by plants and animals to keep gametes isolated and species distinct. Species often have different mating or courtship patterns or behaviours, the breeding seasons may be distinct and even if mating does occur antigenic reactions to the sperm of other species prevent fertilization or embryo development. The Lonicera fly is the first known animal species that resulted from natural hybridization. Until the discovery of the Lonicera fly, this process was known to occur in nature only among plants.

CLONING

Cloning is the process of creating an identical copy of something. In biology, it collectively refers to processes used to create copies of DNA fragments (molecular cloning), cells (cell cloning), or organisms. The term also encompasses situations whereby organisms reproduce asexually.

Etymology

The term *clone* is derived from *êëþí*, the Greek word for "twig, branch", referring to the process whereby a new plant can be created from a twig. In horticulture, the spelling *clon* was used until the twentieth century; the final *e* came into use to indicate the vowel is a "long o" instead of a "short o" [*citation needed*]. Since the term entered the popular lexicon in a more general context, the spelling *clone* has been used exclusively.

Molecular Cloning

Molecular cloning refers to the procedure of isolating a defined DNA sequence and obtaining multiple copies of it *in vivo*. Cloning is frequently employed to amplify DNA fragments containing genes, but it can be used to amplify any DNA sequence such as promoters, non-coding sequences and randomly fragmented DNA. It is utilised in a wide array of biological experiments and practical applications such as large scale protein production. Occasionally, the term cloning is misleadingly used to refer to the identification of the chromosomal location of a gene associated with a particular phenotype of interest, such as in positional cloning. In practice, localization of the gene to a chromosome or genomic region does not necessarily enable one to isolate or amplify the relevant genomic sequence.

In essence, in order to amplify any DNA sequence in a living organism, that sequence must be linked to an origin of replication, a sequence element capable of directing the propagation of itself and any linked sequence. In practice, however, a number of other features are desired and a variety of specialised cloning vectors exist that allow protein expression, tagging, single stranded RNA and DNA production and a host of other manipulations.

Cloning of any DNA fragment essentially involves four steps: fragmentation, ligation, transfection, and screening/selection. Although these steps are invariable among cloning procedures a number of alternative routes can be selected, these are summarised as a 'cloning strategy'.

Initially, the DNA of interest needs to be isolated to provide a relevant DNA segment of suitable size. Subsequently, a ligation procedure is employed whereby the amplified fragment is inserted into a vector. The vector (which is frequently circular) is linearised by means of restriction enzymes, and incubated with the fragment of interest under appropriate conditions with an enzyme called DNA ligase. Following ligation the vector with the insert of interest is transfected into cells. A number of alternative techniques are available, such as chemical sensitivation of cells, electroporation and biolistics. Finally, the transfected cells are cultured. As the aforementioned procedures are of particularly low efficiency, there is a need to identify the cells that have been successfully transfected with the vector construct containing the desired insertion sequence in the required orientation. Modern cloning vectors include selectable antibiotic resistance markers, which allow only cells in which the vector has been transfected, to grow. Additionally, the cloning vectors may contain colour selection markers which provide blue/white screening (á-factor complementation) on X-gal medium. Nevertheless, these selection steps do not absolutely guarantee that the DNA insert is present in the cells obtained. Further investigation of the resulting colonies is required to confirm that cloning was successful. This may be accomplished by means of PCR, restriction fragment analysis and/or DNA sequencing.

Cellular Cloning

Cloning a cell means to derive a population of cells from a single cell. In the case of unicellular organisms such as bacteria and yeast, this process is remarkably simple and essentially only requires the inoculation of the appropriate medium. However, in the case of cell cultures from multi-cellular organisms, cell cloning is an arduous task as these cells will not readily grow in standard media.

A useful tissue culture technique used to clone distinct lineages of cell lines involves the use of cloning rings (cylinders). According to this technique, a single-cell suspension of cells which have been exposed to a mutagenic agent or drug used to drive selection is plated at high dilution to create isolated colonies; each arising from a single and potentially clonally distinct cell. At an early growth stage when colonies consist of only a few of cells, sterile polystyrene rings (cloning rings), which have been dipped in grease are placed over an individual colony and a small amount of trypsin is added. Cloned cells are collected from inside the ring and transferred to a new vessel for further growth.

Organism

Organism cloning refers to the procedure of creating a new multicellular organism, genetically identical to another. In essence this form of cloning is an asexual method of reproduction, where fertilization or inter-gamete contact does not take place. Asexual reproduction is a naturally occurring phenomenon in many species, including most plants (see vegetative reproduction) and some insects.

Horticultural

The term *clone* is used in horticulture to mean all descendants of a single plant, produced by vegetative reproduction or apomixis. Many horticultural plant cultivars are clones, having been derived from a single individual, multiplied by some process other than sexual reproduction. As an example, some European cultivars of grapes represent clones that have been propagated for over two millennia. Other examples are potato and banana. Grafting can be regarded as cloning, since all the shoots and branches coming from the graft are genetically a clone of a single individual, but this particular kind of cloning has not come under ethical scrutiny and is generally treated as an entirely different kind of operation.

Many trees, shrubs, vines, ferns and other herbaceous perennials form clonal colonies. Parts of a large clonal colony often become detached from the parent, termed fragmentation, to form separate individuals. Some plants also form seeds asexually, termed apomixis, e.g. dandelion.

Animals

Clonal derivation exists in nature in some animal species and is referred to as parthenogenesis. An example is the "Little Fire Ant" (*Wasmannia auropunctata*), which is native to Central and South America but has spread throughout many tropical environments.

First Cloned Buffalo

On September 15, 2007, the Philippines announced its development of Southeast Asia's first cloned water buffalo. The Philippine Council for Agriculture, Forestry and Natural Resources Research and Development (PCARRD), under the Department of Science and Technology in Los Baños, Laguna approved this project. The Department of Agriculture's Philippine Carabao Center (PCC) will implement *"Cloning through somatic cell nuclear transfer as a tool for genetic improvement in water buffaloes."* *"Super buffalo calves"* will be produced. There will be no modification or alteration of the genetic materials, as in GMOs (genetically modified organisms).

Therapeutic Cloning

Therapeutic cloning refers to a procedure which produces cells, specific body parts, and organs to be utilized for medical purposes. Although this has only been realized with parts of bladders, early cleavage-stage human embryos have been cloned and this is the subject of much active research. Currently, patients subjected to transplantation are administered immunosuppressive drugs to prevent recognition of the foreign transplant by their immune system and its subsequent rejection. The ability to clonally derive tissues and organs from the patients' own cells would abolish the need for immunosuppressive drugs and would allow the patients to live a life without the potentially serious side-effects of immunosuppressive drugs. More importantly, the ability to clonally derive organs would alleviate the current shortage of transplants and would possibly reduce waiting times for transplants to become available.

Reproductive Cloning

Reproductive cloning uses "somatic cell nuclear transfer" (SCNT) to create animals that are genetically identical. This process entails the transfer of a nucleus from a donor adult cell (somatic cell) to an egg which has no nucleus. If the egg begins to divide normally it is transferred into the uterus of the surrogate mother.

Such clones are not strictly identical since the somatic cells may contain mutations in their nuclear DNA. Additionally, the mitochondria in the cytoplasm also contains DNA and during SCNT this DNA is wholly from the donor egg, thus the mitochondrial genome is not the same as that of the nucleus donor cell from which it was produced. This may have important implications for cross-species nuclear transfer in which nuclear-mitochondrial incompatibilities may lead to death.

Species Cloned

The modern cloning techniques involving nuclear transfer have been successfully performed on several species. Landmark experiments in chronological order:

- Tadpole: (1952) Many scientists questioned whether cloning had actually occurred and unpublished experiments by other labs were not able to reproduce the reported results.
- Carp: (1963) In China, embryologist Tong Dizhou cloned a fish. He published the findings in an obscure Chinese science journal which was never translated into English.
- Sheep: (1996) From early embryonic cells by Steen Willadsen. Megan and Morag cloned from differentiated embryonic cells in June 1995 and Dolly the sheep from a somatic cell in 1997.
- Rhesus Monkey: Tetra (female, January 2000) from embryo splitting
- Cattle: Alpha and Beta (males, 2001) and (2005) Brazil

- Cat: CopyCat "CC" (female, late 2001), Little Nicky, 2004, was the first cat cloned for commercial reasons
- Mule: Idaho Gem, a john mule born 2003-05-04, was the first horse-family clone.
- Horse: Prometea, a Haflinger female born 2003-05-28, was the first horse clone.

For a complete list see: List of animals that have been cloned.

Health Aspects

The success rate of cloning has been low: Dolly the sheep was born after 277 eggs were used to create 29 embryos, which only produced three lambs at birth, only one of which lived. Seventy calves have been created from 9,000 attempts and one third of them died young; Prometea took 328 attempts. Notably, although the first clones were frogs, no adult cloned frog has yet been produced from a somatic adult nucleus donor cell.

There were early claims that Dolly the Sheep had accelerated aging. Aging of this type is thought to be due to the shortening of telomeres, regions at the tips of chromosomes which prevent genetic threads from fraying every time a cell divides. Over time telomeres get worn down until cell-division is no longer possible — this is thought to be a cause of aging. Dolly died in the year 2003. Ian Wilmut said that Dolly's early death had nothing to do with cloning but with a respiratory infection common to lambs raised like Dolly.

Analysis of telomeres from cloned cows showed that they were longer than noncloned calves. This suggests clones could live longer life spans although many died young after excessive growth. Researchers think that this could eventually be developed to reverse aging in humans, provided that aging is based chiefly on the shortening of telomeres. Although some work has been performed on telomeres and aging in nuclear transfer clones, the evidence is at an early stage.

Dolly the Sheep

Dolly (1996-07-05 – 2003-02-14), a Finn Dorsett ewe, was the first mammal to have been successfully cloned from an adult cell. She was cloned at the Roslin Institute in Scotland and lived there until her death when she was six. On 2003-04-09 her stuffed remains were placed at Edinburgh's Royal Museum, part of the National Museums of Scotland.

Dolly was publically significant because the effort showed that the genetic material from a specific adult cell, programmed to express only a distinct subset of its genes, could be reprogrammed to grow an entire new organism. Before this demonstration, there was no proof for the widely spread hypothesis that differentiated animal cells can give rise to entire new organisms.

Human Cloning

Human cloning is the creation of a genetically identical copy of an existing or previously existing human. The term is generally used to refer to *artificial* human cloning; human clones in the form of identical twins are commonplace, with their cloning occurring during the natural process of reproduction. There are two commonly discussed types of human cloning: therapeutic cloning and reproductive cloning. A third type of cloning called replacement cloning exists in theory, and is a combination of therapeutic and reproductive cloning. Replacement cloning entails the replacement of an extensively damaged, failed, or failing body through cloning followed by whole or partial brain transplant. It has been proposed as a way to greatly extend lifespan.

Human cloning is among the most controversial forms of the practice. There have been numerous demands for all progress in the human cloning field to be halted. Some people and groups oppose therapeutic cloning but many more oppose reproductive cloning. The American Association for the Advancement of Science (AAAS) and other scientific organizations have made public statements suggesting that human reproductive cloning be banned until safety issues are resolved. Serious ethical issues have arisen in discussions of harvesting of organs from clones. Some people have considered the idea of growing organs separately from a human organism - in doing this, a new organ supply could be established without the moral implications of harvesting them from human organisms. Research is also being done on the idea of growing organs that are biologically acceptable to the human body inside of other organisms, such as pigs or cows, then transplanting them to humans.

Ethical Issues of Cloning

Although the practice of cloning organisms has been widespread for several thousands of years in the form of horticultural cloning, the recent technological advancements that have allowed for cloning of animals (and potentially humans) have been highly controversial. Some believe it is unethical to use a human clone to save the life of another. Others have countered that people who exist today and have interpersonal relationships and personal histories should take precedence over never-conscious life at any stage of developmental maturity. The Catholic Church and various traditionalist religious groups oppose all forms of cloning, on the grounds that life begins at conception. Conversely, Judaism does not equate life with conception and, though some question the wisdom of cloning, Orthodox rabbis generally find no firm reason in Jewish law and ethics to object to cloning. From the standpoint of classical liberalism, concerns also exist regarding the protection of the identity of the individual and the right to protect one's genetic identity.

The social implications of an artificial human production scheme were famously explored in the novel *Brave New World* and also in the movie *The Island*. These fictional works explore the prejudices that could develop against clones with respect to a kind of slavery or caste system.

Cloning Extinct and Endangered Species

Cloning, or more precisely, the reconstruction of functional DNA from extinct species has, for decades, been a dream of some scientists. The possible implications of this were dramatized in the best-selling novel by Michael Crichton and high budget Hollywood thriller *Jurassic Park*. In real life, one of the most anticipated targets for cloning was once the Woolly Mammoth, but attempts to extract DNA from frozen mammoths have been unsuccessful, though a joint Russo-Japanese team is currently working toward this goal.

In 2001, a cow named Bessie gave birth to a cloned Asian gaur, an endangered species, but the calf died after two days. In 2003, a banteng was successfully cloned, followed by three African wildcats from a thawed frozen embryo. These successes provided hope that similar techniques (using surrogate mothers of another species) might be used to clone extinct species. Anticipating this possibility, tissue samples from the last *bucardo* (Pyrenean Ibex) were frozen immediately after it died. Researchers are also considering cloning endangered species such as the giant panda, ocelot, and cheetah. The "Frozen Zoo" at the San Diego Zoo now stores frozen tissue from the world's rarest and most endangered species.

In 2002, geneticists at the Australian Museum announced that they had replicated DNA of the Thylacine (Tasmanian Tiger), extinct about 65 years previous, using polymerase chain reaction. However, on 2005-02-15 the museum announced that it was stopping the project after tests showed the specimens' DNA had been too badly degraded by the (ethanol) preservative. Most recently, on 2005-05-15, it was announced that the Thylacine project would be revived, with new participation from researchers in New South Wales and Victoria.

One of the continuing obstacles in the attempt to clone extinct species is the need for nearly perfect DNA. Cloning from a single specimen could not create a viable breeding population in sexually reproducing animals. Furthermore, even if males and females were cloned, the question would remain open if they would be viable at all in the absence of parents that could teach or show them their natural behavior. Essentially, if cloning an extinct species succeeded — it must be considered that cloning still is an experimental technology that succeeds only by chance — it is far more likely than not that any resulting animals, even if they were healthy, would be little more than curios or museum pieces.

Cloning endangered species is a highly ideological issue. Many conservation biologists and environmentalists vehemently oppose cloning endangered species — not because they think it won't work but because they think it may deter donations to help preserve natural habitat and wild animal populations. The "rule-of-thumb" in animal conservation is that, if it is still feasible to conserve habitat and viable wild populations, breeding in captivity should not be undertaken in isolation.

In a 2006 review, David Ehrenfeld concludes that cloning in animal conservation is an experimental technology that, at its present state, cannot be expected to work except by pure chance and utterly fails a cost-benefit analysis. Furthermore, he says, it is likely to siphon funds from established and working projects and does not address any of the issues underlying animal extinction (such as habitat destruction, hunting or other overexploitation, and an impoverished gene pool). While cloning technologies are well-established and used on a regular basis in plant conservation, care must be taken to ensure genetic diversity. He concludes:

> "Vertebrate cloning poses little risk to the environment, but it can consume scarce conservation resources, and its chances of success in preserving species seem poor. To date, the conservation benefits of transgenics and vertebrate cloning remain entirely theoretical, but many of the risks are known and documented. Conservation biologists should devote their research and energies to the established methods of conservation, none of which require transgenics or vertebrate cloning."

Embryo

Somatic cell nuclear transfer can also be used to create a clonal embryo. The most likely scenario for this is to produce embryos for use in research, particularly stem cell research. This process is also called "research cloning" or "therapeutic cloning."

Therapeutic cloning, also called "embryo cloning," is the production of human embryos for use in research. The goal of this process is not to create cloned human beings, but rather to harvest stem cells that can be used to study human development and to treat disease. Stem cells are important to biomedical researchers because they can be used to generate virtually any type of specialized cell in the human body. Stem cells are extracted from the egg after it has divided for 5 days. The egg at this stage of development is called a blastocyst. Many researchers hope that one day stem cells can be used to serve as replacement cells to treat heart disease, Alzheimer's, cancer, and other diseases.

Scientists believe that cloning may be used to create stem cells genetically compatible with the somatic cell donor. Cloning in stem cell research, called *research cloning* or *therapeutic cloning*, has not yet been successful: no embryonic stem cell lines have been derived from clonal embryos. The process might provide a way to grow organs in host carriers, so that organs could be produced which would be completely compatible with the original tissue donor. Host carrier growing poses a risk of trans-species diseases if the host is of a different species (e.g., a pig).

In human beings, this is a highly controversial issue for several reasons. It involves creating human embryos in vitro and then destroying them during the process of attempting to obtain embryonic stem cells. But proposals to use cloning techniques in human stem cell research raise a set of concerns beyond the moral status of the embryo. These have led a number of individuals and organizations, who are not opposed in principle to human embryonic stem cell research, to be concerned about or opposed to, human research cloning. One concern is that cloning in human stem cell research will lead to the reproductive cloning of humans. A second concern relates to the appropriate sourcing of the eggs that are needed. Research cloning requires a large number of human eggs, which can only be obtained from women. A third concern is the feasibility of developing stem cell therapies from cloning.

REFERENCES

L.L. Cavalli-Sforza and A.W.F. Edwards (Sep., 1967). "Phylogenetic analysis: Models and estimation procedures". *Evol.* 21 (3): 550-570.

Hardy, G. H. (1908). "Mendelian proportions in a mixed population". *Science* 28: 49 – 50.

Peterson, A.T. and I.L. Brisbin, Jr.: 1999. Genetic endangerment of wild red junglefowl (Gallus gallus) ? Bird Conservation International (Vol. 9) pp. 387–394.

Brisbin, I.L., Jr. (1969). Behavioral differentiation of wildness in two strains of Red Junglefowl (abstract). Amer. Zool. 9:1072.

Manila Times, RP to produce Southeast Asia's first cloned buffalo

Wikinews: Endangered cow cloned in Brazil, 2005-05-22.

Vogel, Gretchen (2000). "In Contrast to Dolly, Cloning Resets Telomere Clock in Cattle". *Science* 288: 641. "analyses indicated that the telomeres were also extended beyond those of newborn".

Pence, Gregory E. (1998). *Who's Afraid of Human Cloning?*. Rowman & Littlefield. paperback.

"Scientists 'to clone mammoth'", BBC News, 2003-08-18.

Heidi B. Perlman. "Scientists Close on Extinct Cloning", Associated Press, 2000-10-08.

Pence, Gregory E. (2005). *Cloning After Dolly: Who's Still Afraid?*. Rowman & Littlefield.

Holloway, Grant. "Cloning to revive extinct species", CNN.com, 2002-05-28.

Ehrenfeld, David (2006). "Transgenics and Vertebrate Cloning as Tools for Species Conservation". *Conservation Biology* 20 (3): 723-732.

5

Overview of Genetic Engineering and Bioengineering

GENETIC ENGINEERING

Genetic engineering, recombinant DNA technology, genetic modification/manipulation (GM) and gene splicing are terms that are applied to the direct manipulation of an organism's genes. Genetic modification of an organism can be achieved through a number of methods, most notably traditional breeding and recombinant technologies. The goal of both is the same, introduction of DNA (in the form of a gene) which in turn finds expression as favorable physical characteristics. Transgenic endeavors have found success in improving crop technology, the manufacture of human insulin through the use of modified bacteria, the manufacture of erythropoietin in Chinese hamster ovary cells, and the production of new types of experimental mice such as the OncoMouse (cancer mouse) for research.

Since a protein sequence is specified by a segment of DNA called a gene, novel versions of that protein can be produced by changing the DNA sequence of the gene. Some groups have argued that genetic engineering is wrong and is "doing the work of God", but most scientists believe that genetic engineering is essential to help future medical discoveries. However, even with regard to this technology's great potential, scientists around the world have raised concerns about the introduction of genetically engineered plants and animals into the environment and the potential dangers of human consumption of GM foods. They say that these organisms have the potential to spread their modified genes into native populations thereby disrupting natural ecosystems., and Genetically Modified Organism (GMO) for more information on GM controversies. Professor Stephen Hawking defended the genetic enhancing of our species in order to compete with Artificial intelligence.

Methodology

There are a number of ways through which genetic engineering is accomplished. Essentially, the process has four main steps.

1. Isolation
2. Insertion into a vector
3. transformation
4. Tests to isolate genetically modified organism (GMO)

Isolation is achieved by identifying the gene of interest that the scientist wishes to insert into the organism, usually using existing knowledge of the various functions of genes. DNA information can be obtained from cDNA or gDNA libraries, and amplified using PCR techniques. If necessary, ie for insertion of eukaryotic genomic DNA into prokaryotes, further modification may be carried out such as removal of introns or ligating prokaryotic promoters.

Once the gene of interest is isolated, it is inserted into a vector such as a plasmid. Other vectors can also be used, such as viral vectors, and non-prokaryotic ones such as liposomes, or even direct insertion using DNA guns. Restriction enzymes and ligases are of great use in this crucial step if it is being inserted into prokaryotic or viral vectors. Daniel Nathans and Hamilton Smith received the 1978 Nobel Prize in Physiology or Medicine for their isolation of restriction endonucleases.

Once the vector is obtained, it can be used to transform the target organism. Depending on the vector used, it can be complex or simple. For example, using raw DNA with DNA guns is a fairly straightforward process but with low success rates, where the DNA is coated onto particles such as gold and fired directly into a cell. Other more complex methods, such as bacterial transformation or using virus's as vectors have higher success rates.

After transformation, the GMO can be isolated from those that have failed to take up the vector in various ways. One method is testing with DNA probes that can hybridize to the gene of interest that was supposed to have been inserted, another would be to package resistance genes along with the vector, such that the resulting GMO is resistant to certain chemicals, such as penicillin, and then they can be grown on agar dishes with penicillin, to ensure only those that have taken up the vector will survive.

Applications

The first genetically engineered drug was human insulin, approved by the United States Food and Drug Administration in 1982. Another early application of genetic engineering was to create human growth hormone as replacement for a drug that was previously extracted from human cadavers. In 1986 the FDA approved the first genetically engineered vaccine for humans, for hepatitis B. Since these early uses of the technology in medicine, the use of GE has expanded to supply many drugs and vaccines.

One of the best known applications of genetic engineering is the creation of genetically modified organisms (GMOs) such as foods and vegetables that resist pest and bacteria infection and have longer freshness.

There are potentially momentous biotechnological applications of GM, for example oral vaccines produced naturally in fruit, at very low cost.

An ambition of some GE is human enhancement via genetics, eventually by molecular engineering.

Genetic Engineering and Research

Although there has been a tremendous revolution in the biological sciences in the past twenty years, there is still a great deal that remains to be discovered. The completion of the sequencing of the human genome, as well as the genomes of most agriculturally and scientifically important plants and animals, has increased the possibilities of genetic research immeasurably. Expedient and inexpensive access to comprehensive genetic data has become a reality with billions of sequenced nucleotides already online and annotated.

- Loss of function, such as in a knockout experiment, in which an organism is engineered to lack the activity of one or more genes. This allows the experimenter to analyze the defects caused by this mutation, and can be considerably useful in unearthing the function of a gene. It is used especially frequently in developmental biology. A knockout experiment involves the creation and manipulation of a DNA construct in vitro, which, in a simple knockout, consists of a copy of the desired gene which has been slightly altered such as to cripple its function. The construct is then taken up by embryonic stem cells, where the engineered copy of the gene replaces the organism's own gene. These stem cells are injected into blastocysts, which are implanted into surrogate mothers. Another method, useful in organisms such as Drosophila (fruitfly), is to induce mutations in a large population and then screen the progeny for the desired mutation. A similar process can be used in both plants and prokaryotes.
- Gain of function experiments, the logical counterpart of knockouts. These are sometimes performed in conjunction with knockout experiments to more

finely establish the function of the desired gene. The process is much the same as that in knockout engineering, except that the construct is designed to increase the function of the gene, usually by providing extra copies of the gene or inducing synthesis of the protein more frequently.

- 'Tracking' experiments, which seek to gain information about the localization and interaction of the desired protein. One way to do this is to replace the wild-type gene with a 'fusion' gene, which is a juxtaposition of the wild-type gene with a reporting element such as Green Fluorescent Protein (GFP) that will allow easy visualization of the products of the genetic modification. While this is a useful technique, the manipulation can destroy the function of the gene, creating secondary effects and possibly calling into question the results of the experiment. More sophisticated techniques are now in development that can track protein products without mitigating their function, such as the addition of small sequences which will serve as binding motifs to monoclonal antibodies.
- Expression studies aim to discover where and when specific proteins are produced. In these experiments the DNA sequence before the DNA that codes for a protein, known as a gene's promoter is reintroduced into an organism with the protein coding region replaced by a reporter gene such as GFP or an enzyme that catalyzes the production of a dye. Thus the time and place where a particular protein is produced can be observed. Expression studies can be taken a step further by altering the promoter to find which pieces are crucial for the proper expression of the gene and are actually bound by transcription factor proteins; this process is known as promoter bashing.

BIOENGINEERING

Biological engineering (also biosystems engineering and bioengineering) deals with engineering biological processes in general. It is a broad-based engineering discipline that also may involve product design, sustainability and analysis of biological systems. In other words, Bioengineering is a discipline that applies engineering principles to biological systems for the purpose of developing new technologies of services to improve the living standards of societies. It exploits the new developments in molecular biology, biochemistry, microbiology, cell metabolism and engineering principles and applies them in order to understand living systems and to bring solutions to various problems associated with these systems.

Bioengineering employs the knowledge and expertise in a number of pure and applied sciences such as; mass and heat transfer, kinetics, biocatalysis, biomechanics, bioinformatics, separation, and purification processes, bioreactor design, surface science, fluid mechanics, thermodynamics, and polymer science constitute the fundamentals of engineering and couples it with knowledge in biological sciences such as genetics, molecular biology, protein chemistry, cytology, neurobiology, immunology and pharmacology, aiming to solve the different problems of the various sectors of the modern societies.

Because other engineering disciplines overlap bioengineering living organisms (e.g., prosthetics in mechanical engineering), the term can be applied more broadly to include food engineering and biotechnology. Biological engineering is called Bioengineering by some colleges and Biomedical engineering is called Bioengineering by others, and is a rapidly developing field with fluid categorization. However, the Main Fields of Bioengineering may be categorised as:

- Bioprocess Engineering; Bioprocess Design, Biocatalysis, Bioseparation Bioinformatics.
- Genetic Engineering; Cell Engineering, Tissue Culture Engineering.
- Biomedical Engineering; Biomedical Diagnosis, Biomedical Therapy, Biomechanics, Biomaterials. Generally, bioengineering may deal with either the medical (see biomedical engineering) or the agricultural fields (see agricultural engineering).

Biological engineers are similar to biologists in that they study living organisms. They are engineers because they have a practical design aim in mind - they use research to create usable tangible products. In general, biological engineers attempt to 1) mimic biological systems in order to create products or 2) modify and control biological systems so that they can replace, augment, or sustain chemical and mechanical processes.

CRYOBIOLOGY

Cryobiology is the study of living organisms, organs, biological tissues or biological cells at low temperatures. This knowledge is practically applied in three fields: cryonics, cryopreservation and cryosurgery.

Definitions/Distinctions

Cryobiology is the branch of biology that studies the effects of low temperatures on organisms (most often for the purpose of achieving cryopreservation). The word cryobiology (from the Greek words "cryo" = cold, "bios" = life, and "logos" = science) literally signifies the science of life at low temperatures. In practice, this field comprises the study of any biological material or system (e.g., proteins, cells, tissues, organs, or organisms) subjected to any temperature below normal (ranging from moderately hypothermic conditions to cryogenic temperatures). At least 6 major areas of cryobiology can be identified: 1) study of cold-adaptation of microorganisms, plants (= cold hardiness), invertebrates, and animals (= hibernation), 2) cryopreservation of cells, tissues, gametes, and embryos of animal and human origin for (medical) purposes of long-term storage. This usually requires the addition of substances which protect the cells during freezing and thawing (cryoprotectants), 3) preservation of organs under hypothermic conditions for transplantation, 4) lyophilization (freeze-drying) of pharmaceuticals, 5) cryosurgery, a (minimally) invasive approach for the destruction of unhealthy tissue using cryogenic gases/fluids, and 6) physics of supercooling, ice nucleation/growth and mechanical engineering aspects of heat transfer during cooling and warming.

Cryogenics

The branch of physics (or engineering) that studies very low temperatures (or the production of very low temperatures). People often confuse "cryogenics" with "cryonics".

Cryonics

Cryonics is the low temperature preservation of humans and animals in a damaged state with the intention of future revival. Cryonics uses information from many fields, including cryobiology, to reduce damage during preservation and measure what preservation is achieved. However no human has ever been revived from cryopreservation by cryonics, and proposed methods for revival are based on speculation concerning capabilities of future technology. Therefore cryonics is viewed with skepticism by most scientists and medical doctors today, although there are some scientists involved in cryonics.

Historical Background

Cryobiology history can be traced back to antiquity. As early as in 2500 BC low temperatures were used in Egypt in medicine. The use of cold was recommended by Hippocrates to stop bleeding and swelling. With the emergence of modern science, Robert Boyle studied the effects of low temperatures on animals.

In 1949 sperm was cryopreserved for the first time by a team of scientists led by Christopher Polge (1926-2006). This led to a much wider use of cryopreservation today, with many organs, tissues and cells routinely stored in low temperatures. Large organs such as hearts are usually stored and transported, for short times only, at cool but not freezing temperatures for transplantation. Cell suspensions (like blood and semen) and thin tissue sections can sometimes be stored almost indefinitely at liquid nitrogen temperature (cryopreservation). Human sperm, eggs and embryos are routinely stored in fertility research and treatments. In the early

2000s a baby was born from a cryopreserved egg fertilized by a cryopreserved sperm.

Cryosurgery (= intended and controlled tissue destruction by ice formation) was carried out by James Arnott in 1845 in an operation on a patient with cancer. Although not very widespread, cryosurgery has its benefits. Hypothermia, e.g. during heart surgery on a "cold" heart (generated by cold perfusion without any ice formation) allows for much longer operations and improves recovery rates for patients.

Applied Cryobiology

Cryobiology as an applied science is primarily concerned with low temperature preservation. Hypothermic storage is typically above 0°C but below normothermic (32°C to 37°C) mammalian temperatures. Storage by cryopreservation, on the other hand, will be in the "80°C to "196°C temperature range. Organs, and tissues are more frequently the objects of hypothermic storage, whereas single cells have been the most common objects cryopreserved.

A rule of thumb in hypothermic storage is that every 10°C reduction in temperature is accompanied by a 50% decrease in oxygen consumption. Although hibernating animals have adapted mechanisms to avoid metabolic imbalances associated with hypothermia, hypothermic organs and tissues being maintained for transplantation require special preservation solutions to counter acidosis, depressed sodium pump activity and increased intracellular calcium. Special organ preservation solutions such as Viaspan (University of Wisconsin solution), HTK, and Celsior have been designed for this purpose. These solutions also contain ingredients to minimize damage by free radicals, prevent edema, compensate for ATP loss, etc.

Cryopreservation of cells is guided by the "Two-Factor Hypothesis" of American cryobiologist Peter Mazur, which states that excessively rapid cooling kills cells by intracellular ice formation and excessively slow cooling kills cells by either electrolyte toxicity or mechanical crushing. During slow cooling ice forms extracellularly, causing water to osmotically leave cells, thereby dehydrating them. Intracellular ice can be much more damaging than extracellular ice.

For red blood cells the optimum cooling rate is very rapid (nearly 100°C per second), whereas for stem cells the optimum cooling rate is very slow (1°C per minute). Cryoprotectants, such as DMSO (dimethyl sulfoxide) and glycerol, are used to protect cells from freezing. A variety of cell types are protected by 10% DMSO. Cryobiologists attempt to optimize cryoprotectant concentration (minimizing both ice formation and toxicity) as well as cooling rate. Cells may be cooled at an optimum cooling rate to a temperature between "30°C and "40°C before being plunged into liquid nitrogen.

Slow cooling methods rely on the fact that cells contain few nucleating agents, but contain naturally-occurring vitrifying substances that can prevent ice formation in cells that have been moderately dehydrated. Cryobiologists are increasingly using mixtures of cryoprotectants for full vitrification (zero ice formation) in preservation of cells, tissues and organs. Vitrification methods pose a challenge in the requirement to search for cryoprotectant mixtures that can minimize toxicity.

Scientific Societies

One of the two leading scientific societies in the field of cryobiology is the Society for Cryobiology. This society was founded in 1964 to bring together those from the biological, medical and physical sciences who have a common interest in the effect of low temperatures on biological systems. As of November 2006 the Society for Cryobiology had approximately 300 members from around the world, and one half of them are US based. The purpose of the Society is to promote scientific research in low temperature biology, to improve scientific understanding in this field, and to disseminate and apply this knowledge to the benefit of mankind. The Society requires of all its members the highest ethical and scientific standards in the performance of their professional activities. According to the Society's Bylaws membership may be refused to applicants whose conduct is deemed detrimental to the Society, this includes explicitly any practice or application of freezing deceased persons in the anticipation of their reanimation. This is the characteristic difference between "cryobiology" (which is a scientific discipline) and "cryonics". The Society organizes an annual scientific meeting dedicated to all aspects of low-temperature biology. This international meeting offers opportunities for presentation and discussion of the most up-to-date research in cryobiology as well as reviewing specific aspects through symposia and workshops. Members are also kept informed of news and forthcoming meetings through the Society newsletter, News Notes. The 2006-2007 President of the Society for Cryobiology is Andreas Sputtek.

The Society for Low Temperature Biology was founded in 1964 and became a Registered Charity in 2003 (Charity Commission for England & Wales No. 1099747) with the purpose of promoting research into the effects of low temperatures on all types of organisms and their constituent cells, tissues and organs. As of 2006 the Society for Low Temperature Biology had approximately 130 (mostly British and European) members and holds at least one Annual General Meeting. The program usually includes both a symposium on a topical subject and a session of free communications on any aspect of low temperature biology. Recent symposia have included long-term stability, preservation of aquatic organisms, cryopreservation of embryos and gametes, preservation of plants, low temperature microscopy, vitrification (glass formation of aqueous systems during cooling), freeze drying and tissue banking. Members are informed through the Society Newsletter, which is presently published

3 times a year. The 2005-2006 Chair of the Society for Low Temperature Biology has been Tiantian Zhang.

A list of additional scientific societies (mostly using "applied" cryobiology) can be found here.

Journals

CRYOBIOLOGY, (publisher: Elsevier) is the foremost scientific publication in this area, with approximately 60 refereed contributions published each year. Articles concern any aspect of low temperature biology and medicine (e.g. freezing, freeze-drying, hibernation, cold tolerance and adaptation, cryoprotective compounds, medical applications of reduced temperature, cryosurgery, hypothermia, and perfusion of organs).

There is an independent UK based rapid communication journal named *CRYO LETTERS* which publishes papers on the effects produced by low temperatures on a wide variety of biophysical and biological processes, or studies involving low temperature techniques in the investigation of biological and ecological topics.

CELL PRESERVATION TECHNOLOGY is a peer-reviewed quarterly scientific journal published by Mary Ann Liebert, Inc. dedicated to the diverse spectrum of preservation technologies including cryopreservation, dry-state (anhydrobiosis), glassy-state and hypothermic maintenance.

MICROBIAL BIODEGRADATION

Interest in the microbial biodegradation of pollutants has intensified in recent years as mankind strives to find sustainable ways to cleanup contaminated environments. These bioremediation and biotransformation methods endeavour to harness the astonishing, naturally occurring, ability of microbial xenobiotic metabolism to degrade, transform or accumulate a huge range of compounds including hydrocarbons (e.g. oil), polychlorinated biphenyls (PCBs), polyaromatic hydrocarbons (PAHs), pharmaceutical substances, radionuclides and metals. Major methodological breakthroughs in recent years have enabled detailed genomic, metagenomic, proteomic, bioinformatic and other high-throughput analyses of environmentally relevant microorganisms providing unprecedented insights into key biodegradative pathways and the ability of organisms to adapt to changing environmental conditions.

The elimination of a wide range of pollutants and wastes from the environment is an absolute requirement to promote a sustainable development of our society with low environmental impact. Biological processes play a major role in the removal of contaminants and they take advantage of the astonishing catabolic versatility of microorganisms to degrade/convert such compounds. New methodological breakthroughs in sequencing, genomics, proteomics, bioinformatics and imaging are producing vast amounts of information. In the field of Environmental Microbiology, genome-based global studies open a new era providing unprecedented *in silico* views of metabolic and regulatory networks, as well as clues to the evolution of degradation pathways and to the molecular adaptation strategies to changing environmental conditions. Functional genomic and metagenomic approaches are increasing our understanding of the relative importance of different pathways and regulatory networks to carbon flux in particular environments and for particular compounds and they will certainly accelerate the development of bioremediation technologies and biotransformation processes.

Aerobic Biodegradation of Pollutants

The burgeoning amount of bacterial genomic data provides unparalleled opportunities for understanding the genetic and molecular bases of the degradation of organic pollutants. Aromatic compounds are among the most recalcitrant of these pollutants and lessons can be learned from the recent genomic studies of *Burkholderia xenovorans* LB400 and *Rhodococcus* sp. strain RHA1, two of the largest bacterial genomes completely sequenced to date. These studies have helped expand our understanding of bacterial catabolism, non-catabolic physiological adaptation to organic compounds, and the evolution of large bacterial genomes. First, the metabolic pathways from phylogenetically diverse isolates are very similar with respect to overall organization. Thus, as originally noted in pseudomonads, a large number of "peripheral aromatic" pathways funnel a range of natural and xenobiotic compounds into a restricted number of "central aromatic" pathways. Nevertheless, these pathways are genetically organized in genus-specific fashions, as exemplified by the b-ketoadipate and Paa pathways. Comparative genomic studies further reveal that some pathways are more widespread than initially thought. Thus, the Box and Paa pathways illustrate the prevalence of non-oxygenolytic ring-cleavage strategies in aerobic aromatic degradation processes. Functional genomic studies have been useful in establishing that even organisms harboring high numbers of homologous enzymes seem to contain few examples of true redundancy. For example, the multiplicity of ring-cleaving dioxygenases in certain rhodococcal isolates may be attributed to the cryptic aromatic catabolism of different terpenoids and steroids. Finally, analyses have indicated that recent genetic flux appears to have played a more significant role in the evolution of some large genomes, such as LB400's, than others. However, the emerging trend is that the large gene repertoires of potent pollutant degraders such as LB400 and RHA1 have evolved principally through more ancient processes. That this is true in such phylogenetically diverse species is remarkable and further suggests the ancient origin of this catabolic capacity.

Anaerobic Biodegradation of Pollutants

Anaerobic microbial mineralization of recalcitrant organic pollutants is of great environmental significance and involves intriguing novel biochemical reactions. In particular, hydrocarbons and halogenated compounds have long been doubted to be degradable in the absence of oxygen, but the isolation of hitherto unknown anaerobic hydrocarbon-degrading and reductively dehalogenating bacteria during the last decades provided ultimate proof for these processes in nature. Many novel biochemical reactions were discovered enabling the respective metabolic pathways, but progress in the molecular understanding of these bacteria was rather slow, since genetic systems are not readily applicable for most of them. However, with the increasing application of genomics in the field of environmental microbiology, a new and promising perspective is now at hand to obtain molecular insights into these new metabolic properties. Several complete genome sequences were determined during the last few years from bacteria capable of anaerobic organic pollutant degradation. The ~4.7 Mb genome of the facultative denitrifying *Aromatoleum aromaticum* strain EbN1 was the first to be determined for an anaerobic hydrocarbon degrader (using toluene or ethylbenzene as substrates). The genome sequence revealed about two dozen gene clusters (including several paralogs) coding for a complex catabolic network for anaerobic and aerobic degradation of aromatic compounds. The genome sequence forms the basis for current detailed studies on regulation of pathways and enzyme structures. Further genomes of anaerobic hydrocarbon degrading bacteria were recently completed for the iron-reducing species *Geobacter metallireducens* (accession nr. NC_007517) and the perchlorate-reducing *Dechloromonas aromatica* (accession nr. NC_007298), but these are not yet evaluated in formal publications. Complete genomes were also determined for bacteria capable of anaerobic degradation of halogenated hydrocarbons by halorespiration: the ~1.4 Mb genomes of *Dehalococcoides ethenogenes* strain 195 and *Dehalococcoides* sp. strain CBDB1 and the ~5.7 Mb genome of *Desulfitobacterium hafniense* strain Y51. Characteristic for all these bacteria is the presence of multiple paralogous genes for reductive dehalogenases, implicating a wider dehalogenating spectrum of the organisms than previously known. Moreover, genome sequences provided unprecedented insights into the evolution of reductive dehalogenation and differing strategies for niche adaptation.

Bioavailability, Chemotaxis, and Transport of Pollutants

Bioavailability, or the amount of a substance that is physiochemically accessible to microorganisms is a key factor in the efficient biodegradation of pollutants. Chemotaxis, or the directed movement of motile organisms towards or away from chemicals in the environment is an important physiological response that may contribute to effective catabolism of molecules in the environment. In addition, mechanisms for the intracellular accumulation of aromatic molecules via various transport mechanisms are also important.

Oil Biodegradation

Petroleum oil contains aromatic compounds that are toxic for most life forms. Episodic and chronic pollution of the environment by oil causes major ecological perturbations. Marine environments are especially vulnerable since oil spills of coastal regions and the open sea are poorly containable and mitigation is difficult. In addition to pollution through human activities, millions of tons of petroleum enter the marine environment every year from natural seepages. Despite its toxicity, a considerable fraction of petroleum oil entering marine systems is eliminated by the hydrocarbon-degrading activities of microbial communities, in particular by a remarkable recently discovered group of specialists, the so-called hydrocarbonoclastic bacteria (HCB) . *Alcanivorax borkumensis* was the first HCB to have its genome sequenced.

Analysis of Waste Biotreatment

Sustainable development requires the promotion of environmental management and a constant search for new technologies to treat vast quantities of wastes generated by increasing anthropogenic activities. Biotreatment, the processing of wastes using living organisms, is an environmentally friendly, relatively simple and cost-effective alternative to physico-chemical clean-up options. Confined environments, such as bioreactors], have been engineered to overcome the physical, chemical and biological limiting factors of biotreatment processes in highly controlled systems. The great versatility in the design of confined environments allows the treatment of a wide range of wastes under optimized conditions. To perform a correct assessment, it is necessary to consider various microorganisms having a variety of genomes and expressed transcripts and proteins. A great number of analyses are often required. Using traditional genomic techniques, such assessments are limited and time-consuming. However, several high-throughput techniques originally developed for medical studies can be applied to assess biotreatment in confined environments.

Metabolic Engineering and Biocatalytic Applications

The study of the fate of persistent organic chemicals in the environment has revealed a large reservoir of enzymatic reactions with a large potential in preparative organic synthesis, which has already been exploited for a number of oxygenases on pilot and even on industrial scale. Novel catalysts can be obtained from metagenomic libraries and DNA sequence based approaches. Our increasing capabilities in adapting the catalysts to specific reactions and process requirements by rational and random mutagenesis broadens the scope for application in the fine chemical industry, but also in the field of biodegradation. In many cases, these catalysts need to be exploited in whole cell bioconversions or in fermentations, calling for system-wide approaches to understanding strain physiology and metabolism and rational approaches to the engineering of whole cells as they are increasingly put forward in the area of systems biotechnology and synthetic biology.

HUMAN GENETIC ENGINEERING

Human genetic engineering refers to the controlled modification of the human genome, which is the genome of *Homo sapiens*, composed of 23 pairs of chromosomes with a total of approximately 3 billion DNA base pairs containing an estimated 30,000 genes. DNA provides the genetic blueprint for all living organisms that influences the physical and mental actions and abilities. With the advent of DNA research and the ability to change gene expressions, it is now possible that scientists may be able to change human capacities, whether they be physical, cognitive, or emotional. Human genetic engineering is still in its infancy, however, with current research generally restricted to animals or gene therapy.

Healthy humans do not need gene therapy to survive (live a socially acceptable lifespan), though it may prove helpful to treat certain diseases. Special gene modification research has been carried out on groups such as the 'bubble children' - those whose immune systems do not protect them from the bacteria and irritants all around them. The first clinical trial of human gene therapy began in 1990, but (as of 2006) is still experimental. Other forms of human genetic engineering are still theoretical, or restricted to fiction stories. Recombinant DNA research is usually performed to study gene expression and various human diseases. Some drastic demonstrations of gene modification have been made with mice and other animals, however; testing on humans is generally considered off-limits. In some instances changes are usually brought about by removing genetic material from one organism and transferring them into another species. This method is known as recombinant genetics.

There are two main types of genetic engineering. Somatic modifications involve adding genes to cells other than egg or sperm cells. For example, if a person had a disease caused by a defective gene, a healthy gene could be added to the affected cells to treat the disorder. The distinguishing characteristic of somatic engineering is that it is non-inheritable, e.g. the new gene would not be passed to the recipient's offspring.

Germline engineering would change genes in eggs, sperm, or very early embryos. This type of engineering is inheritable, meaning that the modified genes would appear not only in any children that resulted from the procedure, but in all succeeding generations. This application is by far the more consequential as it could open the door to the perpetual and irreversible alteration of the human species.

There are two techniques researchers are currently experimenting with:

- Viruses are good at injecting their DNA payload into human cells and reproducing it. By adding the desired DNA to the DNA of non-pathogenic virus, a small amount of virus will reproduce the desired DNA and spread it all over the body.
- Manufacture large quantities of DNA, and somehow package it to induce the target cells to accept it, either as an addition to one of the original 23 chromosomes, or as an independent 24th human artificial chromosome.

Human genetic engineering means that some part of the genes or DNA of a person are changed. It is possible that through engineering, people could be given more arms, bigger brains, or other structural alteration such as fins if desired. A more common type of change would be finding the genes of extraordinary people - such as those for intelligence, stamina, longevity, and incorporating those in embryos. Human genetic engineering holds the promise of being able to cure diseases and increasing the immunity of people to virus. Disorders such as Cystic fibrosis, which is a genetic disease that affects lungs and other organs is the result of small DNA changes.

Researchers are currently trying to map out and assign genes to different body functions or disease. When the genes or DNA sequence responsible for a disease is found, theoretically gene therapy should be able to fix the disease and eliminate it permanently. However, with the complexity of interaction between genes and gene triggers, gene research is currently in its infancy. Computer modeling and expression technology could be used in the future to create people from scratch. This would work by taking existing DNA knowledge and inserting DNA of "superior" body expressions from people, such as a bigger heart, stronger muscles, etc and implanting this within an egg to be inserted into a female womb. The visual modeling of this process may be very much like the videogame "Spore", where people are able to manipulate the physical attributes of creatures and then "release them" in the digital world.

The possibilities of physical changes are endless. Strength, speed, endurance and so on can be enhanced. The baby can be made taller, more beautiful, the changes possible are really up to the imagination, and the ability of the techniques employed by future gene manipulators. Certain people have been identified with extraordinary physical abilities (think athletes, geniuses, physical and mental event record holders) and their genes could be identified and replaced into the target embryo. There is also the possibility that science will advance so much that people will create genes not identified in nature or people and implant those in the human body.

Corresponding gene function to intelligence or mental aptitude in various fields is much harder because while researchers are finding out which sections of the brain light up when used through MRI imaging, corresponding genes to manipulate and/or expand intelligence are harder to map. The brain seems to be the last great medical mystery because unlike a muscle, it transfers information and handles complex processes like a computer but without any logical process discernible to researchers. However, in certain individuals that have a higher aptitude at certain tasks, the history of their family having done the same work seems to show that either through practice, teaching, or gene expressions these individuals find tasks such as composing music or mathematics much easier than the average member of the population.

What Kind of Technology Could be Used for Gene Manipulation?

Current Technology

One available method is gene manipulation through virus insertion. The main mechanism by which this operates is a fairly standard one; one is taking advantage of the way viruses hi-jack the cellular machinery to make multiple copies of themselves. The way they do this is by inserting a segment of their genetic code into the DNA of their host after penetrating the cellular wall, and so use this to instruct their host's machinery to churn out multiple copies of the virus. So, in order to use this to one's advantage, a standard technique might be to modify a single virus' genetic code via certain direct methods, to replace the "hijacking DNA" with the desired sequence, then set the thing loose. Problems with this approach may be:

- Randomness of the insertion of the DNA fragment wanted (this might lead, for instance, to a cell turning cancerous, or being more subtly disrupted),
- Possible lack of control over the immune response to the "retrovirus", and
- Possible problems getting complete transformation of the target cell population.

Another approach is to take advantage of the cell's endogenous ability to undergo homologous recombination wherein an exogenous DNA fragement with ends matching an endogenous gene replaces the target gene by action of enzymes called recombinases. This is the technology used for "knockouts" and "knock-ins" and generally cannot be performed on an intact organism and is therefore only a candidate for germline (i.e. single cell) genetic manipulation. This technology has very widespread usage in mouse genetics wherein genetically altered embryos are grown in a surrogate mother. This could easily be extended to humans in the very near future as there are no major additional technological hurdles remaining to engineer humans versus mice (though minor hurdles certainly remain). The primary hurdle is ethical, as the efficiency of this process can be quite poor and many embryos would be sacrificed in order to produce one genetically modified offspring.

Future Technology

There are limitations to what current technology can do. It is desirable to get finer control on how DNA sequences might be added to a host cell. Technologies that could offer this sort of control might well result from a merger of highly engineered/ modified viruses with micromechanical technologies, modified to migrate to a particular specified N nucleotide sequence in the host DNA for N significantly large to minimize targeting errors, then precisely and efficiently insert the desired snippet in the correct position in the host. Evidently it would be desirable to allow such a device/ organism to feed on available resources in the host in order to migrate from cell to cell and perform this process until a significant proportion of host cells had been transformed; it would then be desirable for the device to disable itself. Also it would be desirable for the device to understand when it had finished with a cell (eg after levels of a desired protein reach a critical threshold, though this would be perhaps too slow, ie movement via measurement of some sort of chemical gradient) so that it could move on. Certainly such devices could not be user controlled for a large population; they would need to be autonomous.

Since DNA sequences vary from individual to individual, and individuals may even be chimeras, also required would be the capability to sequence individual genomes, the ability to target fixes to particular parts of a chimera population, and finally a good understanding of how to avoid disruptive edits- that is, edits that do not disrupt the functioning of some other process in the cell. This requires some level of mastery of the area of proteomics, that is, the understanding of a lifeform's proteome.

But if the genotypes dont match it doesn't necessarily mean that they will be the same in clones. Like for example i had a cat that was aging, and when he was about to die I cloned him and the cat had a different pheno type, but it was nothing like my first cat. We had to put the cat down because the "copycat" had sever disfunctions in his respritory system. Now that I think about it I should have never cloned my animal, just to put it through all that pain.

Computers

The future most likely lies in a process that is automated by a wide repository of knowledge, a method of combining that knowledge into an embryo, and a design system such as a computer program that is easily represented to human individuals. The researcher is able to pick and choose what characteristics the "soon to be created human" should have, such as high intelligence, high strength, 20/20 or above vision, etc, and then tells the computer to create that person. This designer baby machine would then search through current research on these genes, put them into the proper places of the DNA mode through an accurate DNA modification method, enter it into an egg (created or taken from a female), and then insert the egg into a female or a mechanical womb.

Nanotechnology

Nanotechnology could be used to implement gene changes. Nanotechnology is an expression of ultra small devices or machines that are able to operate at the nanoscale level. These could theoretically modify the human body on as directed to make changes.

When to Make Changes

Changes at Conception

Genetic engineering is most easily accomplished by making changes just after the egg and sperm have melded but before first division. In this way, the gene will be expressed throughout and will affect the recipients children, grandchildren, and all subsequent generations. Germline engineering is controversial because there is insufficient knowledge about DNA expression to accurately judge what result these changes will have.

Changes After Birth

As of now, this is likely to take the form of gene therapy. This would not be hereditary unless the sex cells are engineered.

Applications

Curing Medical Conditions

When treating problems that arise from genetic disorder, one solution is gene therapy. A genetic disorder is a situation where some genes are missing or faulty. When this happens, genes may be expressed in unfavorable ways or not at all, and this generally leads to further complications.

The idea of gene therapy is that a non-pathogenic virus or other delivery system can be used to insert a piece of DNA—a good copy of the gene—into cells of the living individual. The modified cells would divide as normal and each division would produce cells that express the desired trait. The result would be that he/she would then have the ability to express the trait that was previously absent at least partially. This form of genetic engineering could help alleviate many problems, such as diabetes, cystic fibrosis, or other genetic diseases.

Human Enhancement

The potential of genetic engineering to cure medical conditions opens the question of exactly what such a condition is. Some view ageing and death as medical conditions and therefore potential targets for engineering solutions. They see human genetic engineering potentially as a key tool in this. The difference between cure and enhancement from this perspective is merely one of degree. Theoretically genetic engineering could be used to drastically change people's genomes, which could enable people to regrow limbs and other organs, perhaps even extremely complex ones such as the spine. It could also be used to make people stronger, faster, smarter, or to increase the capacity of the lungs, among other things. If a gene exists in nature, it could be brought over to a human cell. In this view, there is no qualitative difference (only a quantitative one) between, for instance, a genetic intervention to cure muscular dystrophy, and a genetic intervention to improve muscle function even when those muscles are functioning at or around the human average (since there is also an average muscle function for those with a particular type of dystrophy, which the treatment would improve upon).

Others feel that there is an important distinction between using genetic technologies to treat those who are suffering and to make those who are already healthy superior to the average. There is widespread agreement that germline engineering should not currently be allowed for either therapeutic and enhancement applications, as evidenced by a recent report by the American Association for the Advancement of Science. Though theory and speculation suggest that genetic engineering could be used to make people stronger, faster, smarter, or to increase lung capacity, the AAAS report finds that there is little evidence that this can currently be done without very unsafe and therefore unethical human experiments. Because different cells have different tasks, changing one cell to do a different job will not only affect that one task, it can affect many others too.

Considerations

Ethical Considerations

- We could choose to have changes made to us, but we might also be making the choice for our children if the changes are carried through to the germline. Do we have that right, and how far should we take our ability?
- Can the slippery slope argument be used to defend a prohibition of germ-line gene therapy?
- Conversely, is it responsible and ethically acceptable to leave the potential of our children to the chance effects of the "genetic lottery", if we obtain the technological capacity to make positive changes?

- If genetic engineering became the way of the future, would people whose parents could not afford to genetically 'modify' them while still in an embryo, have a chance of achieving with high standards compared to the people who were 'modified' to be perfect? Should government pay for genetic engineering to prevent this inequality?
- Is it ethical to experiment on embryos that have yet to be born?
- How would genetic engineering be used to revolutionize warfare?
- Does superior ability breed a superior ambition?
- Can we and should we eliminate the flaws in human nature that cause violence and conflict through genetic engineering?
- In gene therapy, does the government have a right to limit what people do with their own bodies?
- In combining human genes and non-human genes, and modifying non-humans with human genes, the question comes about what makes something human or not?
- Is it sheer madness to arrogantly assume that we can improve the handiwork of God and/or evolution?
- If gene therapy can ever prevent aging, how far does the morale obligation to save lives extend?
- If a person's genetics are ever drastically altered, should they still be considered the same person?

Social Considerations

Individuals may benefit from non-therapeutic genetic engineering, but some claim that there may be adverse *social* implications. Few resources – particularly those related to medicine and health care – are available to everyone, and allowing the most privileged to engineer themselves or their children to have special capabilities could lead to what some call a genetic aristocracy. Numerous enhancements via genetic engineering have been proposed, including increased memory, intelligence, and less need for sleep, in addition to some peoples' desires to alter their physical appearance. The advantages created by genetic engineering, either real or perceived, could lead to new forms of inequality between those with genetic enhancements and those without while also exacerbating current inequalities between the rich and poor.

This isn't to say that the poor would never have access to such technologies - there is no theoretical limit on how much retrovirus can be cultured in a bioreactor. But the likely high costs would probably require harsh financing plans. Genetic mortgages of the future could go hand in hand with student loans, spelling disaster for dropouts and increasing the risks of a first-class education for a middle or lower class citizen. Many people view genetic engineering as just a way to cure diseases.

On the other hand society could get to the point where the government pays for everyone to undergo genetic engineering. There would also be a point where patents on specific treatments run out, allowing cheaper access to the majority population after a decade or two of high cost. This may actually decrease inequalities, because natural differences in ability would be nearly eliminated leaving socioeconomic success or failure entirely to the environment and personal choice.

Metaphysical Considerations

- The metaphysical (or "spiritual") implications of genetically engineered humans are vast in scope; e.g., were individual personality shown to be exclusively the result of genetic information acted upon by the environment, the concepts of the human soul and free will could be proven specious.
- However, could this be shown by the success of actual attempts at genetic engineering any more (or less) than it is shown by what is already known about the roles of genes and the environment in shaping the phenotypical characteristics (including behavior) of living things?

REFERENCES

Diaz E (editor). (2008). *Microbial Biodegradation: Genomics and Molecular Biology*, 1st ed., Caister Academic Press.

Heider J and Rabus R (2008). "Genomic Insights in the Anaerobic Biodegradation of Organic Pollutants", *Microbial Biodegradation: Genomics and Molecular Biology*. Caister Academic Press.

Mark S. Frankel & Audrey R. Chapman, Human Inheritable Genetic Modifications: Assessing Scientific, Ethical, Religious, and Policy Issues, (September 2000).

Martins VAP et al (2008). "Genomic Insights into Oil Biodegradation in Marine Systems", *Microbial Biodegradation: Genomics and Molecular Biology*. Caister Academic Press.

McLeod MP and Eltis LD (2008). "Genomic Insights Into the Aerobic Pathways for Degradation of Organic Pollutants", *Microbial Biodegradation: Genomics and Molecular Biology*. Caister Academic Press.

Meyer A and Panke S (2008). "Genomics in Metabolic Engineering and Biocatalytic Applications of the Pollutant Degradation Machinery", *Microbial Biodegradation: Genomics and Molecular Biology*. Caister Academic Press.

Parales RE et al (2008). "Bioavailability, Chemotaxis, and Transport of Organic Pollutants", *Microbial Biodegradation: Genomics and Molecular Biology*. Caister Academic Press

Watanabe K and Kasai Y (2008). "Emerging Technologies to Analyze Natural Attenuation and Bioremediation", *Microbial Biodegradation: Genomics and Molecular Biology*. Caister Academic Press.

6

GMOs With Special Focus on Genetically Modified Food

GENETICALLY MODIFIED ORGANISM

A genetically modified organism (GMO) or genetically engineered organism (GEO) is an organism whose genetic material has been altered using the genetic engineering techniques generally known as recombinant DNA technology. With recombinant DNA technology, DNA molecules from different sources are combined *in vitro* into one molecule to create a new gene. This modified DNA is then transferred into an organism causing the expression of modified or novel traits.

The term "GMO" has historically been defined as organisms whose genetic makeup has been altered by conventional cross breeding or by "mutagenesis" breeding, as these methods predate the discovery of the recombinant DNA techniques. However, this term is now interchangeable with Genetically Engineered Organism.

History

The general principle of producing a GMO is to add genetic material into an organism's genome to generate new traits - Genetic engineering - was made possible through a series of scientific advances including the discovery of DNA and the creation of the first recombinant bacteria in 1973, i.e., *E .coli* expressing a salmonella gene. This led to concerns in the scientific community about potential risks from genetic engineering which have been thoroughly discussed at the Asilomar Conference in Pacific Grove, California. The recommendations laid out from this meeting were that government oversight of recombinant DNA research should be established until the technology was deemed safe. Herbert Boyer then founded the first company to use recombinant DNA technology, Genentech, and in 1978 the company announced the creation of an *E. coli* strain producing the human protein insulin.

In 1986, field tests of a bacterium genetically engineered to protect plants from frost damage (ice-minus bacteria) at a small biotechnology company called Advanced Genetic Sciences of Oakland, California, were repeatedly delayed by opponents of biotechnology. In the same year, a proposed field test of a microbe genetically engineered for a pest resistance protein by Monsanto was dropped.

Uses of GMOs

Examples of GMOs are highly diverse, and include transgenic (genetically modified by recombinant DNA methods) animals such as mice, fish, transgenic plants, or various microbes, such as fungi and bacteria. The generation and use of GMOs has many reasons, chief among them are their use in research that addresses fundamental or applied questions in biology or medicine, for the production of pharmaceuticals and industrial enzymes, and for direct, and often controversial, applications aimed at improving human health (e.g., gene therapy) or agriculture (e.g., golden rice). The term "genetically modified organism" does not always imply, but can include, targeted insertions of genes from one into another species. For example, a gene from a jellyfish, encoding a fluorescent protein called GFP, can be physically linked and thus co-expressed with mammalian genes to identify the location of the protein encoded by the GFP-tagged gene in the mammalian cell. These and other methods are useful and indispensable tools for biologists in many areas of research, including those that study the mechanisms of human and other diseases or fundamental biological processes in eukaryotic or prokaryotic cells.

Transgenic Animals

Transgenic animals are used as experimental models to perform phenotypic tests with genes whose function is unknown or to generate animals that are susceptible to certain compounds or stresses for testing in biomedical research. Other applications include the production of human hormones, such as insulin.

Frequently used in genetic research are transgenic fruit flies as genetic models to study the effects of genetic changes on development. Flies are often preferred over other animals for ethical reasons and ease of culture, and also because the fly genome is somewhat simpler than that of vertebrates.

- Enviropig

Transgenic Plants

Transgenic plants have been developed for various purposes: resistance to pests, herbicides or harsh environmental conditions; improved shelflife; increased nutritional value - and many more. Since the first commercial cultivation of GM plants in 1996, GM plant events tolerant to the herbicides glufosinate or glyphosate and events producing the Bt toxin, an insecticide, have dominated the market. Recently, a new generation of GM plants promising benefits for consumers and industry purposes is becoming ready to enter the markets.

Since GM plants are grown on open fields, there is often a perception that there could be associated environmental risks. Therefore, most countries require biosafety studies prior to the approval of a new GM plant event, usually followed by a monitoring programme to detect environmental impacts.

Especially in Europe, the coexistence of GM plants with conventional and organic crops has raised many concerns. Since there is separate legislation for GM crops and a high demand from consumers for the freedom of choice between GM and non-GM foods, measures are required to separate GM, conventional and organic plants and derived food and feed. European research programmes such as Co-Extra, Transcontainer and SIGMEA are investigating appropriate tools and rules. On the field level, these are biological containment methods, isolation distances and pollen barriers. Further down the food chain, documentation and detection methods shall ensure the coexistence.

Controversy Over GMOs

Government Support for and Ban of GMOs

The use of GMOs has sparked significant controversy in many areas. Some groups or individuals see the generation and use of GMO as intolerable meddling with biological states or processes that have naturally evolved over long periods of time, while others are concerned about the limitations of modern science to fully comprehend all of the potential negative ramifications of genetic manipulation.

While some groups advocate the complete prohibition of GMOs, others call for mandatory labeling of genetically modified food or other products. Other controversies include the definition of patent and property pertaining to products of genetic engineering and the possibility of unforeseen local and global effects as a result of transgenic organisms proliferating. The basic ethical issues involved in genetic research are discussed in the article on genetic engineering.

In 2004, Mendocino County, California became the first county in the United States to ban the production of GMOs. The measure passed with a 57% majority. In 2005, a standing committee of the government of Prince Edward Island in Canada began work to assess a proposal to ban the production of GMOs in the province. PEI has already banned GM potatoes, which account for most of its crop. In California, Trinity and Marin counties have also imposed bans on GM crops, while ordinances to do so were unsuccessful in Butte, San Luis Obispo, Humboldt, and Sonoma counties. Supervisors in the agriculturally-rich counties of Fresno, Kern, Kings, Solano, Sutter, and Tulare have passed resolutions supporting the practice.

Currently, there is little international consensus regarding the acceptability and effective role of modified "complete" organisms such as plants or animals. A great deal of the modern research that is illuminating complex biochemical processes and disease mechanisms makes vast use of genetic engineering.

The practice of genetic modification as a scientific technique is not restricted in the United States. Individual genetically modified crops (such as soybeans) are studied before being brought to market, but generally only by the companies providing the modification. This "test by those being tested" practice is common in the United States, where many in the FDA are ex-employees of Monsanto, the largest gene-manipulation firm. Most countries in Europe, Japan, Mexico (among others) have taken the opposite position, stating that genetic modification has not been proven safe, and therefore that they will not accept genetically modified food from the United States or any other country without assessing their safety themselves. This issue has been brought before the World Trade Organization, which determined that not allowing GMOs into the country creates an unnecessary obstacle to international trade. Consequently, genetic modification within agriculture is an issue of some strong debate in the United States, the European Union, and some other countries.

Crosspollination Concerns

Some critics have raised the concern that conventionally bred crop plants can be cross-pollinated (bred) from the pollen of modified plants. Pollen can be dispersed over large areas by wind, animals, and insects. Recent research with creeping bentgrass has lent support to the concern when modified genes were found in normal grass up to 21 km (13 miles) away from the source, and also within close relatives of the same genus (*Agrostis*). GM proponents point out that outcrossing, as this process is known, is not new. The same thing happens with any new open-pollinated crop variety—newly introduced traits can potentially cross out into neighbouring crop plants of the same species and, in some cases, to closely related wild relatives. Defenders of GM technology point out that each GM crop is assessed on a case by case basis to determine if there is any risk associated with the outcrossing of the GM trait into wild plant populations. The fact that a GM plant may outcross with a related wild relative is not, in itself, a risk unless such an occurrence has consequences. If, for example, a herbicide resistance trait was to cross into a wild relative of a crop plant it can be predicted that this would not have any consequences except in areas where herbicides are sprayed, such as a farm. In such a setting the farmer can manage this risk by rotating herbicides.

The European Union funds research programmes such as Co-Extra, that investigate options and technologies on the coexistence of GM and conventional farming. This also includes research on biological containment strategies and other measures, that prevent outcrossing and enable the implementation of coexistence.

If patented genes are outcrossed, even accidentally, to other commercial fields and a person deliberately selects the outcrossed plants for subsequent planting then the patent holder has the right to control the use of those crops. This was supported in Canadian law in the case of Monsanto Canada Inc. v. Schmeiser.

'Terminator' and 'Traitor'

An often cited controversy is a hypothetical "Technology Protection" technology dubbed 'Terminator'. This yet-to-be-commercialised technology would allow the production of first generation crops that would not generate seeds in the second generation because the plants yield sterile seeds. The patent for this so-called "terminator" gene technology is owned by Delta and Pine Land and the United States Department of Agriculture. Delta and Pine Land was bought by Monsanto in August 2006. Similarly, the hypothetical Trait-specific Genetic Use Restriction Technology, also known as 'Traitor' or 'T-gut', requires yearly application of a chemical to genetically-modified crops to reactivate engineered traits. This technology is intended both to limit the spread of genetically engineered plants, and to require farmers to pay yearly to reactivate the genetically engineered traits of their crops. Traitor is under development by companies including Monsanto and AstraZeneca.

In addition to the commercial protection of proprietary technology in selfpollinating crops such as soybean (a generally contentious issue) another purpose of the terminator gene is to prevent the escape of genetically modified traits from crosspollinating crops into wild-type species by sterilizing any resultant hybrids. The terminator gene technology created a backlash amongst those who felt the technology would prevent re-use of seed by farmers growing such terminator varieties in the developing world and was ostensibly a means to exercise patent claims. Use of the terminator technology would also prevent "volunteers", or crops that grow from unharvested seed, a major concern that arose during the Starlink debacle.There are technologies evolving which contain the transgene by biological means and still can provide fertile seeds using fertility restorer functions. Such methods are being developed by several EU research programmes, among them Transcontainer and Co-Extra.

Losses of Biodiversity

In agriculture and animal husbandry, the green revolution popularized the use of conventional hybridization to increase yield many folds by creating "high-yielding varieties". Often the handful of breeds of plants and animals hybridized originated in developed countries and were further hybridized with local varieties, in the rest of the developing world, to create high yield strains resistant to local climate and diseases. Local governments and industry since have been pushing hybridization with such zeal that several of the wild and indigenous breeds evolved locally over thousands of years having high resistance to local extremes in climate and immunity to diseases etc. have already become extinct or are in grave danger of becoming so in the near future. Due to complete disuse because of un-profitability and uncontrolled crosspollination and crossbreeding (genetic pollution) formerly huge gene pools of various wild and indigenous breeds have collapsed causing widespread genetic erosion and genetic pollution resulting in great loss in genetic diversity and biodiversity as a whole.

Concerns have been raised that Genetically Modified Organisms may contribute to genetic pollution because artificially created and genetically engineered plants and animals in laboratories, which could never have evolved in nature even with conventional hybridization, can interbreed with naturally evolved wild varieties. Genetically Modified (GM) crops today have become a common source for genetic pollution, not only of wild varieties but also of other domesticated varieties derived from conventional hybridization.

It is being said that genetic erosion coupled with genetic pollution is destroying that needed unique genetic base thereby creating an unforeseen hidden crisis which will result in a severe threat to our food security for the future when diverse genetic material will cease to exist to be able to further improve or hybridize weakening food crops and livestock against more resistant diseases and climatic changes.

GENETICALLY MODIFIED FOOD

Genetically modified (GM) foods are foodstuffs produced from genetically modified organisms (GMO) that have had their genome altered through genetic engineering. GM Foods have been available since the 1990s. The most common modified foods are derived from plants: soybean, corn, canola and cotton seed oil and wheat.

The process of producing a GMO used for GM Foods may involve taking DNA from one organism, modifying it in a laboratory, and then inserting it into the target organism's genome to produce new and useful traits or phenotypes. Such GMOs are generally referred to as transgenics. Other methods of producing a GMO include increasing or decreasing the number of copies of a gene already present in the target organism, silencing or removing a particular gene or modifying the position of a gene within the genome.

Controversies surrounding GM foods and crops commonly focus on human and environmental safety, labeling and consumer choice, intellectual property rights, ethics, food security, poverty reduction, and environmental conservation.

The second commercially grown genetically modified food crop was the Flavr Savr tomato, which was made more resistant to rotting by Californian company Calgene. Calgene was allowed to release the tomatoes into the market in 1994 without any special labeling. It was welcomed by consumers that purchased the fruit at two to five times the price of regular tomatoes. However, production problems and competition from a conventionally bred, longer shelf-life variety prevented the product from becoming profitable. A variant of the Flavr Savr was used by Zeneca to produce tomato paste which was sold in Europe during the summer of 1996. The labeling and pricing were designed as a marketing experiment, which proved, at the time, that European consumers would accept genetically engineered foods.

The attitude toward GM foods would be drastically changed after outbreaks of Mad Cow Disease weakened consumer trust in government regulators, and protesters rallied against the introduction of Monsanto's "Roundup-Ready" soybeans. The next GM crops included insect-protected cotton and herbicide-tolerant soybeans both of which were commercially released in 1996. GM crops have been widely adopted in the United States. They have also been extensively planted in several other countries (Argentina, Brazil, South Africa, India, and China) where agriculture is a major part of the total economy. Other GM crops include insect-protected maize and herbicide-tolerant maize, cotton, and rapeseed varieties.

Abundance of GM Crops

Between 1996 and 2005, the total surface area of land cultivated with GMOs had increased by a factor of 50, from 17,000 km^2 (4.2 million acres) to 900,000 km^2 (222 million acres), of which 55 percent were in the United States.

Although most GM crops are grown in North America, in recent years there has been rapid growth in the area sown in developing countries. For instance in 2005 the largest increase in crop area planted to GM crops (soybeans) was in Brazil (94,000 km^2 in 2005 versus 50,000 km^2 in 2004.) There has also been rapid and continuing expansion of GM cotton varieties in India since 2002. (Cotton is a major source of vegetable cooking oil and animal feed.) It is predicted that in 2006/7 32,000 km^2 of GM cotton will be harvested in India (up more than 100 percent from the previous season). Indian national average cotton yields of GM cotton were seven times lower in 2002, because the parental cotton plant used in the genetic engineered was not well suited to the climate of India and failed. The publicity given to transgenic trait Bt insect resistance has encouraged the adoption of better performing hybrid cotton varieties, and the Bt trait has substantially reduced losses to insect predation. Economic and environmental benefits of GM cotton in India to the individual farmer have been documented.

In 2003, countries that grew 99 percent of the global transgenic crops were the United States (63 percent), Argentina (21 percent), Canada (6 percent), Brazil (4 percent), China (4 percent), and South Africa (1 percent). The Grocery Manufacturers of America estimate that 75 percent of all processed foods in the U.S. contain a GM ingredient. In particular, Bt corn, which produces the pesticide within the plant itself is widely grown, as are soybeans genetically designed to tolerate glyphosate herbicides. These constitute "input-traits" are aimed to financially benefit the producers, have indirect environmental benefits and marginal cost benefits to consumers.

In the US, by 2006 89% of the planted area of soybeans, 83 percent of cotton, and 61 percent maize was genetically modified varieties. Genetically modified soybeans carried herbicide tolerant traits only, but maize and cotton carried both herbicide tolerance and insect protection traits (the latter largely the *Bacillus thuringiensus* Bt insecticidal protein). In the period 2002 to 2006, there were significant increases in the area planted to Bt protected cotton and maize, and herbicide tolerant maize also increased in sown area.

Future Developments

Future envisaged applications of GMOs are diverse and include drugs in food, bananas that produce human vaccines against infectious diseases such as Hepatitis B, metabolically engineered fish that mature more quickly, fruit and nut trees that yield years earlier, and plants that produce new plastics with unique properties. While their practicality or efficacy in commercial production has yet to be fully tested, the next decade may see exponential increases in GM product development as researchers gain increasing access to genomic resources that are applicable to organisms beyond the scope of individual projects. Safety testing of these products will also at the same time be necessary to ensure that the perceived benefits will indeed outweigh the perceived and hidden costs of development. Plant scientists, backed by results of modern comprehensive profiling of crop composition, point out that crops modified using GM techniques are less likely to have unintended changes than are conventionally bred crops.

Intellectual Property

Monsanto Canada Inc. v. Schmeiser

Enforcement of patents on genetically modified plants is often contentious, especially because of the occurrence of Gene flow. In 1998, 95-98 percent of about 10 km^2 planted with canola by Canadian farmer Percy Schmeiser were found to contain

Monsanto's patented Roundup Ready gene although Schmeiser had never purchased seed from Monsanto. The initial source of the plants was undetermined, and could have been through either gene flow or intentional theft. However, the overwhelming predominance of the trait implied that Schmeiser must have intentionally selected for it. The court determined that Schmeiser had saved seed from areas on and adjacent to his property where Roundup had been sprayed, such as ditches and near power poles.

Although unable to prove direct theft, Monsanto sued Schmeiser for piracy since he knowingly grew Roundup Ready plants without paying royalties(Ibid). The case made it to the Canadian Supreme Court, which in 2004 ruled 5 to 4 in Monsanto's favor. The dissenting judges focused primarily on the fact that Monsanto's patents covered only the gene itself and glyphosate resistant *cells*, and failed to cover transgenic plants in their entirety.

In response to criticism, Monsanto Canada's director of public affairs stated that "It is not, nor has it ever been Monsanto Canada's policy to enforce its patent on Roundup Ready crops when they are present on a farmer's field by accident...Only when there has been a knowing and deliberate violation of its patent rights will Monsanto act." Currently Percy Schmeiser spends a large amount of his time traveling and speaking about how Monsanto ruined his career as a farmer. He also talks about the possible harms of genetically modification and why others in addition to himself should be protesting it.

Coexistence and Traceability

In many countries, and especially in the European Union, consumers demand the choice between foods of genetically modified, conventional or organic origins. This requires a labelling system as well as the reliable separation of GM and non-GM organisms at production level and throughout the whole processing chain.

Research has demonstrated, that coexistence of GM crops can be realised by several agricultural measures, such as isolation distances or biological containment strategies.

For traceability, the OECD has introduced a "unique identifier" which is given to any GMO when it is approved. This unique identifier must be forwarded at every stage of processing.

Many countries have established labelling regulations and guidelines on coexistence and traceability. Research projects such as Co-Extra, SIGMEA and Transcontainer are aimed at investigating improved methods for ensuring coexistence and providing stakeholders the tools required for the implementation of coexistence and traceability.

Policy Around the World

Some argue that there is more than enough food in the world and that the hunger crisis is caused by problems in food distribution and politics, not production, so people should not be offered food that may carry some degree of risk.

Others oppose genetic engineering on the grounds that genetic modifications might have unforeseen consequences, both in the initially modified organisms and their environments. For example, certain strains of maize have been developed that are toxic to plant eating insects (see Bt corn). It has been alleged those strains cross-pollinated with other varieties of wild and domestic maize and passed on these genes with a putative impact on Maize biodiversity. Subsequent to the publication of these results, several scientists pointed out that the conclusions were based on experiments with design flaws. It is well known that the results from the Polymerase Chain Reaction method of analysing DNA can often be confounded by sample contamination and experimental artifacts. Appropriate controls can be included in experiments to eliminate these as a possible explanation of the results - however these controls were not included in the methods used by Quist and Chapela. After this criticism *Nature*, the scientific journal where this data was originally published concluded that "the evidence available is not sufficient to justify the publication of the original paper". More recent attempts to replicate the original studies have concluded that genetically modified corn is absent from southern Mexico in 2003 and 2004 Also in dispute is the impact on biodiversity of the introgression of transgenes into wild populations. Unless a transgene offers a massive selective advantage in a wild population, a transgene that enters such a population will be maintained at a low gene frequency. In such situations it can be argued that such an introgression actually *increases* biodiversity rather than lowers it.

Activists opposed to genetic engineering say that with current recombinant technology there is no way to ensure that genetically modified organisms will remain under control, and the use of this technology outside secure laboratory environments carries potentially unacceptable risks to both farmed and wild ecosystems.

Potential impact on biodiversity may occur if herbicide-tolerant crops are sprayed with herbicide to the extent that no wild plants ('weeds') are able to survive. Plants toxic to insects may mean insect-free crops. This could result in declines in other wildlife (e.g. birds) which feed on weed seeds and/or insects for food resources. The recent (2003) farm scale studies in the UK found this to be the case with GM sugar beet and GM rapeseed, but not with GM maize (though in the last instance, the non-GM comparison maize crop had also been treated with environmentally-damaging pesticides subsequently (2004) withdrawn from use in the EU).

Although some scientists have claimed that selective breeding is a form of genetic engineering, (e.g., maize was modified from teosinte, dogs have evolved with human intervention over the course of tens of thousands of years from wolves), others assert that modern transgenesis-based genetic engineering is capable of delivering changes faster than, and sometimes of different types from, traditional breeding methods.

Proponents of current genetic techniques as applied to food plants cite the benefits that the technology can have, for example, in the harsh agricultural conditions of Africa. They say that with modifications, existing crops would be able to thrive under the relatively hostile conditions providing much needed food to their people. Proponents also cite golden rice and golden rice 2, genetically engineered rice varieties (still under development) that contain elevated vitamin A levels. There is hope that this rice may alleviate vitamin A deficiency that contributes to the death of millions and permanent blindness of 500,000 annually.

Proponents say that genetically-engineered crops are not significantly different from those modified by nature or humans in the past, and are as safe or even safer than such methods. There is gene transfer between unicellular eukaryotes and prokaryotes. There have been no known genetic catastrophes as a result of this. They argue that animal husbandry and crop breeding are also forms of genetic engineering that use artificial selection instead of modern genetic modification techniques. It is politics, they argue, not economics or science, that causes their work to be closely investigated, and for different standards to apply to it than those applied to other forms of agricultural technology.

Proponents also note that species or general barriers have been crossed in nature in the past. An oft-cited example is today's modern red wheat variety, which is the result of two natural crossings made long ago. It is made up of three groups of seven chromosomes. Each of those three groups came from a different wild wheat grass. First, a cross between two of the grasses occurred, creating the durum wheats, which were the commercial grains of the first civilizations up through the Roman Republic. Then a cross occurred between that 14-chromosome durum wheat and another wild grass to create what became modern red wheat at the time of the Roman Empire.

Economic and Political Effects

- Many opponents of current genetic engineering believe the increasing use of GM in major crops has caused a power shift in agriculture towards Biotechnology companies, which are gaining more control over the production chain of crops and food, and over the farmers that use their products, as well.

- Many proponents of some current genetic engineering techniques believe it will lower pesticide usage and has brought higher yields and profitability to many farmers, including those in developing nations. A few genetic engineering licenses allow farmers in less economically developed countries to save seeds for next year's planting.
- In August 2002, Zambia cut off the flow of Genetically Modified Food (mostly maize) from UN's World Food Programme. This left a famine-stricken population without food aid.
- In December 2005 the Zambian government changed its mind in the face of further famine and allowed the importation of GM maize. However, the Zambian Minister for Agriculture Mundia Sikatana has insisted that the ban on genetically modified maize remains, saying "We do not want GM (genetically modified) foods and our hope is that all of us can continue to produce non-GM foods."
- In April 2004 Hugo Chávez announced a total ban on genetically modified seeds in Venezuela.
- In January 2005, the Hungarian government announced a ban on importing and planting of genetic modified maize seeds, although these were agreed authorized by the EU.
- On August 18, 2006, American exports of rice to Europe were interrupted when much of the U.S. crop was confirmed to be contaminated with unapproved engineered genes, possibly due to accidental cross-pollination with conventional crops. The U.S. government has since declared the rice safe for human consumption, and exports to some countries have since resumed.

Conventional Hybridization for Higher Yield, Genetic Engineering and the Resulting loss of Biodiversity, a Threat to Food Security

In agriculture and animal husbandry, green revolution popularized the use of conventional hybridization to increase yield many folds by creating "high-yielding varieties". Often the handful of breeds of plants and animals hybridized originated in developed countries and were further hybridized with local verities, in the rest of the developing world, to create high yield strains resistant to local climate and diseases. Local governments and industry since have been pushing hybridization with such zeal that several of the wild and indigenous breeds evolved locally over thousands of years having high resistance to local extremes in climate and immunity to diseases etc. have already become extinct or are in grave danger of becoming so in the near future. Due to complete disuse because of un-profitability and uncontrolled intentional, compounded with unintentional crosspollination and crossbreeding (genetic pollution) formerly huge gene pools of various wild and indigenous breeds have collapsed causing widespread genetic erosion and genetic pollution resulting in great loss in genetic diversity and biodiversity as a whole.

A Genetically Modified Organism (GMO) is an organism whose genetic material has been altered using the genetic engineering techniques generally known as recombinant DNA technology. Genetic Engineering today has become another serious and alarming cause of genetic pollution because artificially created and genetically engineered plants and animals in laboratories, which could never have evolved in nature even with conventional hybridization, can live and breed on their own and what is even more alarming interbreed with naturally evolved wild varieties. Genetically Modified (GM) crops today have become a common source for genetic pollution, not only of wild varieties but also of other domesticated varieties derived from relatively natural hybridization.

It is being said that genetic erosion coupled with genetic pollution is destroying that needed unique genetic base thereby creating an unforeseen hidden crisis which will result in a severe threat to our food security for the future when diverse genetic material will cease to exist to be able to further improve or hybridize weakening food crops and livestock against more resistant diseases and climatic changes.

CANOLA

Canola is a type of edible oil. It is a trademarked quality description of a group of cultivars of rapeseed variants from which low erucic acid rapeseed oil and low glucosinolate meal are obtained. Also known as "LEAR" oil (for Low Erucic Acid Rapeseed), Canola was initially bred in Canada by Keith Downey and Baldur Stefansson in the 1970s.

The word "canola" was derived from "Canadian oil, low acid" in 1978.

History

Once considered a specialty crop in Canada, canola has evolved into a major North American cash crop. Canada and the United States produce between 7 and 10 million metric tons (tonnes) of canola seed per year. Annual Canadian exports total 3 to 4 million metric tons of the seed, 700,000 metric tons of canola oil and 1 million metric tons of canola meal. The United States is a net consumer of canola oil. The major customers of canola seed are Japan, Mexico, China and Pakistan, while the bulk of canola oil and meal goes to the United States, with smaller amounts shipped to Taiwan, Mexico, China, and Europe. World production of rapeseed oil in the 2002–2003 season was about 14 million metric tons.

Canola was developed through conventional plant breeding from rapeseed, an oilseed plant with roots in ancient civilization. The word "rape" in rapeseed comes from the Latin word "*rapum*," meaning turnip. Turnip, rutabaga, cabbage, Brussels sprouts, mustard and many other vegetables are related to the two canola species commonly grown: *Brassica napus* and *Brassica rapa*. The negative associations with the word "rape" in North America resulted in the more marketing-friendly name "Canola". The change in name also serves to distinguish it from regular rapeseed oil, which has much higher erucic acid content.

Hundreds of years ago, Asians and Europeans used rapeseed oil in lamps. As time progressed, people employed it as a cooking oil and added it to foods. Its use was limited until the development of steam power, when machinists found rapeseed oil clung to water- and steam-washed metal surfaces better than other lubricants. World War II saw high demand for the oil as a lubricant for the rapidly increasing number of steam engines in naval and merchant ships. When the war blocked European and Asian sources of rapeseed oil, a critical shortage developed and Canada began to expand its limited rapeseed production.

After the war, demand declined sharply and farmers began to look for other uses for the plant and its products. Edible rapeseed oil extracts were first put on the market in 1956–1957, but these suffered from several unacceptable characteristics. Rapeseed oil had a distinctive taste and a disagreeable greenish colour due to the presence of chlorophyll. It also contained a high concentration of erucic acid. Experiments on animals have pointed to the possibility that erucic acid, consumed in large quantities, may cause heart damage, though Indian researchers have published findings that call into question these conclusions and the implication that the consumption of mustard or rapeseed oil is dangerous. Feed meal from the rapeseed plant was not particularly appealing to livestock, due to high levels of sharp-tasting compounds called glucosinolates.

Rapeseed had been grown in Canada (mainly Saskatchewan) since 1936. Canadian plant breeders took up the challenge to improve the quality of the plant. In 1968, Dr. Baldur Stefansson of the University of Manitoba used selective breeding to develop a low erucic acid variety of rapeseed. In 1974 another variety was produced with both a low erucic acid content and a low level of glucosinolates; this was dubbed Canola, from Canadian Oil Low Acid.

A variety developed in 1998 is considered to be the most disease- and drought-resistant variety of Canola to date. Recent varieties such as this have been produced by gene splicing techniques.

Canola could emerge as a crop in Central Oregon. An Oregon State University researcher has determined that growing winter canola for hybrid seed now appears possible. Canola is the highest producing oil-seed crop, but the state prohibits it from being grown in Deschutes, Jefferson and Crook counties because it may attract bees away from specialty seed crops such as carrots which require bees for pollination.

Health Effects

Canola oil has been touted as a healthy oil due to its low saturated fat and high monounsaturated oil content—the latter almost 60%—and beneficial omega-3 fatty acids profile. The Canola Council of Canada states it is completely safe and is the healthiest of all commonly used cooking oils. Traditional rapeseed oil contains higher amounts of erucic acid and glucosinolates, both of which were deemed undesirable for human consumption by the United States Food and Drug Administration (FDA). Erucic acid is implicated with cancer and rancidity and glucosinolates are goitrogenic. Canola oil reduces them to very low levels—0.5 to 1% for erucic acid—which is below the 2 percent limit set by the USDA.

For many years, rapeseed oil was used for human consumption in Canada. Although the undesirable effects of glucosinolates and erucic acid were known, they were deemed an acceptable risk versus the many health benefits of rapeseed oil. Nonetheless, researchers attempted and were able to develop fully "double-zero" varieties by the 1980s without significant levels of those two compounds.

Nonetheless, the oil generated controversy with an article implicating Canola oil with glaucoma and the Mad Cow Disease. This article was taken up, condensed and widely circulated in a story via emails. The industry and many health professionals condemn this as an email hoax and condemn its claims as being wholly unsubstantiated.

Genetic Modification

Genetically modified canola which is resistant to herbicide was first introduced to Canada in 1995. Today 80% of acres sown to canola are sown with genetically modified canola.

Introduction of the genetically modified crop to Australia generated considerable controversy. Canola is Australia's third biggest crop, and is often used by wheat farmers as a break crop to improve soil quality. As at 2003 the only genetically modified crops in Australia were non-food crops: carnations and cotton. In 2003, Australia's gene technology regulator approved the release of canola altered to make it resistant to the herbicide Glufosinate ammonium.

REFERENCES

Anderson, K. and Lee Ann Jackson. 2005. Some Implications of GM Food Technology Policies for Sub-Saharan Africa. Journal of African Economies 14(3):385-410;

Heong, KL, YH Chen, DE Johnson, GC Jahn, M Hossain, RS Hamilton. 2005. Debate Over a GM Rice Trial in China. Letters. Science, Vol 310, Issue 5746, 231-233 , 14 October 2005.

Huang, J., Ruifa Hu, Scott Rozelle, Carl Pray. 2005. Insect-Resistant GM Rice in Farmers' Fields: Assessing Productivity and Health Effects in China. Science (29 April 2005) Vol. 308. no. 5722, pp. 688 – 690. DOI: 10.1126/science.1108972

Plague (1978) aka M3: The Gemini Strain, discussion of actors, plot, and critique of the Canadian genetically-modified organism escape thriller.

Martineau, Belinda (2001). *First Fruit: The Creation of the Flavr Savr Tomato and the Birth of Biotech Foods*. McGraw-Hill, 269.

Munzer, Stephen R. (2006). *"Plants, Torts, and Intellecutal Property"*. Oxford University Press.

Federal court of Canada. Monsanto Canada Inc. v. Schmeiser Date: 20010329 Docket: T-1593-98 Retrieved 26-Mar-2006.

Schubert, Robert: "Schmeiser Wants to Take It to The Supreme Court", CropChoice News, Sept. 9, 2002

Quist D and Chapela IH (2001) "Transgenic DNA introgressed into traditional maize landraces in Oaxaca, Mexico". *Nature* 414 (6863): 541-543.

Christou, Paul (2002). "No Credible Scientific Evidence is Presented to Support Claims that Transgenic DNA was Introgressed into Traditional Maize Landraces in Oaxaca, Mexico". *Transgenic Research* 11 (1): 3-5.

"Biodiversity (Communications arising): Suspect evidence of transgenic contamination". *Nature* 416 (6881): 600-601.

Fedoroff, Nina V. (2003) "Prehistoric GM corn". *Science* 302: 1158-1159.

Schubert D (2005) "Regulatory regimes for transgenic crops". *Nat Biotechnol* 23: 785-787.

Ghafoorunissa (1996). "Fats in Indian Diets and Their Nutritional and health Implications". *Lipids* 31: S287-S291.

Shenolikar, I (1980). "Fatty Acid Profile of Myocardial Lipid in Populations Consuming Different Dietary Fats". *Lipids* 15(11): 980-982.

Bellenand, JF; Baloutch, G; Ong, N; Lecerf, J (1980). "Effects of Coconut Oil on Heart Lipids and on Fatty Acid Utilization in Rapeseed Oil". *Lipids* 15(11): 938-943.

Achaya, KT (1987). "Fat Status of Indians - A Review". *Journal of Scientific & Industrial Research* 46: 112-126.

Indu, M; Ghafoorunissa (1992). "n-3 Fatty Acids in Indian Diets - Comparison of the Effects of Precursor (Alpha-Linolenic Acid) Vs Product (Long chain n-3 Poly Unsaturated Fatty Acids)". *Nutrition Research* 12: 569-582.

Glossary

Abiotic stress The effect of non-living factors which can harm living organisms. These non-living factors include drought, extreme temperatures, pollutants, etc.

Adenovirus A group of DNA viruses which cause diseases in animals. In man, they produce acute respiratory tract infections with symptoms resembling common cold. They are used in gene cloning, as vectors for expressing large amounts of recombinant proteins in animal cells. They are also used to make live-virus vaccines against more dangerous pathogens. *See* viral vaccines.

ADEPT (antibody-directed enzyme pro-drug therapy) A way to target a drug to a specific tissue. The drug is administered as an inactive pro-drug, and then converted into an active drug by an enzyme administered with a second injection. The enzyme is coupled to an antibody that concentrates it in the target tissue. When the enzyme arrives at the target tissue, the pro-drug is activated to form the active drug, while elsewhere it remains inactive. *See* drug delivery; targeted drug delivery.

Adhesion (L. *adhaerere*, to stick to) The attraction of dissimilar molecules for each other. A sticking together of unlike substances, such as soil and water.

A-DNA A right-handed DNA double helix that has 11 base pairs per turn. DNA exists in this form when partially dehydrated.

Adsorption The formation of a layer of gas, liquid or solid on the surface of a solid. *cf* absorption.

Adult cloning The creation of identical copies of an adult animal by nuclear transfer (q.v.) from differentiated adult tissue. *See also* cloning; Dolly.

Advanced Applied to an organism or a part thereof, implying considerable development from the ancestral stage or from the explant stage.

Adventitious (L. *adventitius*, not properly belonging to) A structure arising at sites other than the usual ones, e.g., shoots from roots or leaves, and embryos from any cell other than a zygote.

Aerate To supply with or mix with air or gas. The process is aeration.

Aerobe A micro-organism that grows in the presence of oxygen. Opposite: anaerobe.

Aerobic bacteria Bacteria that can live in the presence of oxygen.

Aerobic respiration A type of respiration in which foodstuffs are completely oxidized to carbon dioxide and water, with the release of chemical energy, in a process requiring atmospheric oxygen.

Aerobic Active in the presence of free oxygen.

Affinity chromatography A method for separating molecules by exploiting their ability to bind specifically to other molecules. There are several types of biological affinity chromatography. A biological molecule can be immobilized and a smaller molecule (ligand, q.v.,) to which it is to bind can be stuck to it, or the smaller ligand can be immobilized and the macromolecule stuck to it. A variant is to use an antibody as the immobilized molecule and use it to "capture" its antigen: this is often called **immuno-affinity chromatography**. A variation is **pseudo-affinity chromatography**, in which a compound which is like a biological ligand is immobilized on a solid material, and enzymes or other proteins are bound to it. Other techniques include **metal affinity chromatography**, where a metal ion is immobilized on a solid support: metal ions bind tightly and specifically to many biomolecules. The metal ion is bound to a chelator or chelating group, a chemical group that binds specifically and extremely tightly to that metal.

Affinity tag; purification tag An amino acid sequence that has been engineered into a protein to make its purification easier. These can work in a number of ways. The tag could be another protein, which binds to some other material very tightly and thus allowing the protein to be purified by affinity chromatography (q.v.). The tag could be a short amino acid sequence, which is recognized by an antibody. The antibody would then bind to the protein whereas it would not have done so before. One such short peptide, called FLAG, has been designed so that it is particularly easy to make antibodies against it. The tag could be a few amino acids, which are then used as a chemical tag on the protein. For example, a string of positively charged amino acids will bind very strongly to a negatively charged filter: this could be used as the basis of a separation system. Some amino acids bind metals very strongly, especially in pairs: this chemical property can be exploited by using a filter with metal atoms chemically linked onto it to pull a protein out of a mixture of proteins. *cf* affinity chromatography.

Aflatoxin Toxic compounds, produced by moulds (fungi) of the *Aspergillus flavus* group, that bind to DNA and prevent replication and transcription. Aflatoxins can cause acute liver damage and cancer. Animals may be poisoned by eating stored food or feed contaminated with the mould.

Agar (Malay, *agar-agar*) A polysaccharide solidifying agent used in nutrient media preparations and obtained from certain types of red algae (*Rhodophyta*). Both the type of agar and its concentration can affect the growth and appearance of cultured explants.

Agarose The main constituent of agar.

Agarose gel electrophoresis A process in which a matrix composed of a highly purified form of agar is used to separate larger DNA and RNA molecules. *See* electrophoresis.

Agonist A drug, hormone or transmitter substance that forms a complex with a receptor site that is capable of triggering an active response from a cell.

Agrobacterium A genus of bacteria that includes several plant pathogenic species, causing tumour-like symptoms. *See Agrobacterium tumefaciens;* crown gall; hairy root culture; Ri plasmid; Ti plasmid.

Agrobacterium tumefaciens A bacterium that causes crown gall disease in some plants. The bacterium infects a wound, and injects a short stretch of DNA into some of the cells around the wound. The DNA comes from a large plasmid - the Ti (tumour induction) plasmid - a short region of which (called T-DNA, = transferred DNA) is transferred to the plant cell, where it causes the cell to grow into a tumour-like structure. The T-DNA contains genes which *inter alia* allows the infected plant cells to make two unusual compounds, nopaline and octopine, that are characteristic of transformed cells. The cells form a gall, which hosts the bacterium. This DNA-transfer mechanism is exploited in the genetic engineering of plants. The Ti plasmid is modified so that a foreign gene is transferred into the plant cell along with or instead of the nopaline synthesis genes. When the bacterium is cultured with isolated plant cells or with wounded plant tissues, the "new" gene is injected into the cells and ends up integrated into the chromosomes of the plant.

***Agrobacterium tumefaciens*-mediated transformation** A naturally occurring process of DNA transfer from the bacterium *A. tumefaciens* to plants.

Agrobiodiversity; agricultural biological diversity That component of biodiversity that is relevant to food and agriculture production. The term agrobiodiversity encompasses within-species, species and ecosystem diversity. (Based on FAO, 1999)

AIDS (acquired immunodeficiency syndrome) The usually fatal human disease in which the immune system is destroyed by a retrovirus (Human Immunodeficiency Virus, HIV). The virus infects and destroys helper T-cells, which are essential for combating infections.

Airlift fermenter A cylindrical fermentation vessel in which the cells are mixed by air introduced at the base of the vessel and that rises through the column of culture medium. The cell suspension circulates around the column as a consequence of the gradient of air bubbles in different parts of the reactor.

Albinism Hereditary absence of pigment in an organism. Albino animals have no colour in their skin, hair and eyes. The term is also used for absence of chlorophyll in plants.

Algal biomass Single-celled plants, such as *Chlorella* spp. and *Spirulina* spp., are grown commercially in ponds to make feed materials. Chlorella is grown commercially to make into fish food: it is fed to zooplankton, and these in turn are harvested as feed for fish farms. This is a means of converting sunlight into food in a way more convenient and controllable than normal farming.

Alginate Polysaccharide gelling agent.

Alkylating agents Chemicals that transfer alkyl (methyl, ethyl, etc.) groups to the bases in DNA.

Allele (Gr. *allelon*, of one another, mutually each other); **allelomorph** (adj: allelic, allelomorphic). One of a pair, or series, of variant forms of a gene that occur at a given locus in a chromosome. Alleles are symbolized with the same basic symbol (e.g., B for dominant and b for recessive); B_1, B_2, ..., B_n *for n* additive alleles at a locus). In a normal diploid cell there are two alleles of any one gene (one from each parent), which occupy the same relative position (locus) on homologous chromosomes. Within a population there may be more than two alleles of a gene. *See* multiple alleles.

Allele frequency The number of copies of an allele in a population, expressed as a proportion of the total number of copies of all alleles at a locus in a population.

Allele-specific amplification (ASA) The use of polymerase chain reaction (PCR) at a sufficiently high stringency that only a primer with exactly the same sequence as the target DNA will be amplified. A powerful means of genotyping for single-locus disorders that have been characterized at the molecular level.

Allelic exclusion A phenomenon whereby only one functional allele of an antibody gene can be assembled in a given B lymphocyte. The "allele" on the other homologous chromosome in a diploid mammalian cell cannot undergo a functional re-arrangement, which would result in the production of two different antibodies by a single plasma cell.

Allelopathy The phenomenon by which the secretion of chemicals, such as phenolic and terpenoid compounds, by a plant inhibits the growth or reproduction of other plant species with which it is competing.

Allergen An antigen that provokes an immune response.

Allogamy Cross fertilization in plants. *See* fertilization.

Allometric When the growth rate of one part of an organism differs from that of another part or of the rest of the body.

Allopatric speciation Speciation occurring at least in part because of geographic isolation.

Allopolyploid (Gr. *allos*, other, + polyploidy). A polyploid organism (usually a plant) having multiple sets of chromosomes derived from different species. Hybrids are usually sterile, because they do not have sets of homologous chromosomes and therefore pairing cannot take place. However, if doubling of the chromosome number occurs in a hybrid derived from two diploid (2n) species, the resulting tetraploid (4n) is a fertile plant, since it contains two sets of homologous chromosome and pairing may occur; this tetraploid is an allotetraploid.

Allosteric enzyme An enzyme that has two structurally distinct forms, one of which is active and the other inactive. Active forms of allosteric enzymes tend to catalyse the initial step in a pathway leading to the synthesis of molecules. The end product of this synthesis can act as a feedback inhibitor, converting the enzyme to the inactive form, thus controlling the amount of product synthesized.

Allosteric regulation A catalysis-regulating process in which the binding of a small effector molecule to one site on an enzyme affects the activity at another site.

Allosteric transition A reversible interaction of a small molecule with a protein molecule, resulting in a change in the shape of the protein and a consequent alteration of the interaction of that protein with a third molecule.

Allotetraploid An organism with four genomes derived from hybridization of different species. Usually, in forms that become established, two of the four genomes are from one species and two are from another species. *See* allopolyploid.

Allozygote A diploid individual that is homozygous at a locus in which the two genes are not identical by descent from a common ancestor.

Alphalactalbumin Protein component of milk.

Alternative mRNA splicing The inclusion or exclusion of different exons to form different mRNA transcripts. *See* RNA.

Alu sequences A family of 300-bp sequences occurring nearly a million times in the human genome.

Ambient temperature Air temperature at a given time and place; not radiant temperature.

Amino acid (Gr. *Ammon*, from the Egyptian sun god, in M. L. used in connection with ammonium salts). An acid containing the group NH_2. In particular, any of 20 basic building blocks of proteins with a free amino (NH_2) and a free carboxyl (COOH) group, and having the basic formula NH_2 - C_R - COOH. According to the side group R, they are subdivided into: polar or hydrophilic (serine, threonine, tyrosine, asparagine and glutamine); non-polar or hydrophobic (glycine, alanine, valine, leucine, isoleucine, proline, phenylalanine, tryptophan and cysteine);

acidic (aspartic acid and glutamic acid) and basics (lysine, arginine, hystidine). The sequence of amino acids determines the shape, properties and the biological role of a protein. Plants and many micro-organisms can synthesize amino acids from simple inorganic compounds, but animals are unable to synthesize some of them, called essential amino acids, so they must be present in the diet.

Aminoacyl site; A-site One of two sites on ribosomes to which the incoming aminoacyl tRNA binds.

Aminoacyl tRNA synthetase Enzyme that attaches each amino acid to its specific tRNA molecule.

Amitosis Cell division (cytokinesis), including nuclear division through constriction of the nucleus, without chromosome differentiation as in mitosis. The maintenance of genetic integrity and diploidy during amitosis is uncertain. This process occurs in the endosperm of flowering plants.

Amniocentesis A procedure for obtaining amniotic fluid from a pregnant mammal for the diagnosis of some diseases in the unborn foetus. Cells are cultured, and metaphase chromosomes are examined for irregularities (e.g., Down syndrome, spina bifida, etc., in humans).

Amnion The thin membrane that lines the fluid-filled sac in which the embryo develops in higher vertebrates, reptiles and birds.

Amniotic fluid Liquid contents of the amniotic sac of higher vertebrates, containing cells of the embryo (not of the mother). Both fluid and cells are used for diagnosis of genetic abnormalities in the embryo or foetus.

Amorph; null mutation A mutation that obliterates gene function.

Amphidiploid A species or type of plant derived from doubling the chromosomes in the F_1 hybrid of two species; an allopolyploid. In an amphidiploid the two species are known, whereas in other allopolyploids they may not be known.

Amphimixis True sexual reproduction involving the fusion of male and female gametes and the formation of a zygote.

Amplify To increase the number of copies of a DNA sequence, either *in vivo* by inserting into a cloning vector that replicates within a host cell, or *in vitro* by polymerase chain reaction (PCR).

Amylase A group of enzymes that degrade starch, glycogen and other polysaccharides, producing a mixture of glucose and maltose. Plants have both a- and b-amylase; animal have only a-amylase.

Amylolytic The capability of breaking down starch into sugars.

Amylopectin A polysaccharide comprising highly branched chains of glucose molecules. The water-insoluble portion of starch.

Amylose A polysaccharide consisting of linear chains of 100 to 1000 glucose molecules. The water-soluble portion of starch.

Anabolic pathway A pathway by which a metabolite is synthesized; a biosynthetic pathway.

Anaerobe An organism that can grow in the absence of oxygen. Opposite: aerobe.

Anaerobic An environment or condition in which molecular oxygen is not available for chemical, physical or biological processes.

Anaerobic digestion Digestion of materials in the absence of oxygen. *See* anaerobic respiration.

Anaerobic respiration Respiration in which foodstuffs are partially oxidized, with the release of chemical energy, in a process not involving atmospheric oxygen, such as alcoholic fermentation, in which one of the end products is ethanol.

Analogous Features of organisms or molecules that are superficially or functionally similar but have evolved in a different way or contain different compounds.

Anaphase (Gr. *ana*, up + *phais*, appearance) The stage of mitosis or meiosis during which the daughter chromosomes (sister chromatids) pass from the equatorial plate to opposite poles of the cell (toward the ends of the spindle). Anaphase follows metaphase and precedes telophase.

Anchor gene A gene that has been positioned on both the physical map and the linkage map of a chromosome.

Androgen Any hormone that stimulates the development of male secondary sexual characteristics, and contributes to the control of sexual activity in vertebrate animals. Usually synthesized by the testes.

Androgenesis Male parthenogenesis, i.e., the development of a haploid embryo from a male nucleus. The maternal nucleus is eliminated or inactivated subsequent to fertilization of the ovum, and the haploid individual (referred to as androgenetic) contains in its cells the genome of the male gamete only. Androgenesis is detected by cytological staining. *See* anther culture; gynogenesis; parthenogenesis.

Aneuploid (Gr. *aneu*, without + ploid) An organism or cell having a chromosome number that is not an exact multiple of the monoploid *(x)* with one chromosome being present in greater (e.g., trisomic $2n + 1$) or lesser (e.g., monosomic $2n - 1$) number than the normal diploid number.

Animal cell immobilization Entrapment of animal cells in some solid material in order to produce some natural product or genetically engineered protein. Animal cells have the advantage that they already produce many proteins of pharmacological interest, and that genetically engineered proteins are produced by them with the post-translation modifications normal to animals. However, because animal cells are much more fragile than bacterial ones, they cannot tolerate a commercial fermentation process. Typical materials are hollow fibre membrane bioreactors, or porous carriers made of polysaccharide, protein, plastic or ceramic materials with microscopic holes inside which the cells grow.

Animal genetic resources databank A databank that contains inventories of farm animal genetic resources and their immediate wild relatives, including any information that helps to characterize these resources. (Source: FAO, 1999)

Animal genome (gene) bank A planned and managed repository containing animal genetic resources. Repositories include the environment in which the genetic resource has developed, or is now normally found *(in situ)* or facilities elsewhere *(ex situ - in vivo* or *in vitro).* For *in vitro, ex situ* genome bank facilities, germplasm is stored in the form of one or more of the following: semen, ova, embryos and tissue samples. (Source: FAO, 1999)

Anion A negatively charged ion; opposite: cation.

Anneal The pairing of complementary DNA or RNA sequences, via hydrogen bonding, to form a double-stranded polynucleotide. Most often used to describe the binding of a short primer or probe.

Annealing The process of heating (de-naturing step) and slowly cooling (re-naturing step) double-stranded DNA to allow the formation of hybrid DNA or complementary strands of DNA or of DNA and RNA.

Anther culture The aseptic culture of anthers for the production of haploid plants from microspores. *See* androgenesis; gynogenesis; parthenogenesis.

Anther Microsporangium bearing microspores which develop into pollen (microgametophytes). The upper part of a stamen, containing pollen sacs within which are numerous pollen grains.

Anthesis The flowering period or efflorescence. Anthesis is the time of full bloom, which lasts till fruit set.

Anthocyanin Water-soluble blue, purple and red flavonoid pigments found in vacuoles of cells.

Antiauxin A chemical that interferes with the auxin response. Antiauxin may or may not involve prevention of auxin transport or movement in plants. Some antiauxins are said to promote morphogenesis *in vitro,* such as 2,3,5-tri-iodobenzoate (TIBA; f.w. 499.81), or 2,4,5-trichlorophenoxyacetate (2,4,5-T; m.w. 255.49), which stimulate the growth of some cultures.

Antibiosis The prevention of growth or development of an organism by a substance or another organism.

Antibiotic A class of natural and synthetic compounds that inhibit the growth of or kill some micro-organisms. Antibiotics such as penicillin are often used to control (to some extent kill) contaminating organisms. However, resistance to particular antibiotics can be acquired through mutations. Some contaminating organisms are only suppressed or their metabolism slowed to an insignificant level. *See* antibiotic resistance; bactericide; bacteriostat.

Antibiotic resistance The ability of a micro-organism to produce a protein that disables an antibiotic or prevents transport of the antibiotic into the cell.

Antibody (Gr. *anti*, against + body) An immunological protein (called an immunoglobulin, Ig) produced by certain white blood cells (lymphocytes) of the immune system of an organism in response to a contact with a foreign substance (antigen). Such an immunological protein has the ability of specifically binding with the foreign substance and rendering it harmless. The basic immunoglobulin molecule consists of two identical heavy and two identical light chains. *See* monoclonal antibodies; polyclonal antibodies.

Antibody class The class to which an antibody belongs, depending on the type of heavy chain present. In mammals, there are five classes of antibodies: IgA, IgD, IgE, IgG, and IgM.

Antibody structure Antibodies have a well-defined structure. Each antibody has two identical "light" chains and two identical "heavy" chains. Each chain comprises a constant region, i.e., a region that is the same between antibodies of the same class and sub-class, and a variable region that differs between. The antigen-binding region or binding site - *complementarity determining region* - is in the variable region. The antibody can be cut by proteases into several fragments, known as Fab, Fab', and Fc.

Antibody-mediated (humoral) immune response The synthesis of antibodies by B cells in response to an encounter of the cells of the immune system with a foreign immunogen.

Anticlinal The plane of cell wall orientation or cell division perpendicular (at right angles) to the surface of an organ. *See* tunica; periclinal.

Anticoding strand The strand of the DNA double helix that is actually transcribed. Also known as the antisense or template strand.

Anticodon A triplet of nucleotides in a tRNA molecule that pairs with a complementary triplet of nucleotides, or codon, in an mRNA molecule during translation. *See* codon; mRNA; RNA; tRNA.

Antigen; immunogen A compound that elicits an immune response by stimulating the production of antibodies. The antigen, usually a protein, when introduced into a vertebrate organism is bound by the antibody or a T cell receptor. *See* antigenic determinant; antigenic switching.

Antigenic determinant A surface feature of a micro-organism or macromolecule, such as a glycoprotein, that elicits an immune response. *See* epitope.

Antigenic switching The altering of a micro-organism's surface antigens through genetic re-arrangement, to elude detection by the host's immune system.

Antihaemophilic globulin Blood globulin that reduces the clotting time of haemophilic blood.

Anti-idiotype antibodies Antibodies which recognize the binding sites of other antibodies. Their binding sites are complementary to the binding sites of another immunoglobulin. When an animal becomes immune to something, it not only acquires antibodies against that something, it also acquires antibodies against those antibodies. This forms a network of antibodies which can all bind to each other to various degrees, helping to regulate the immune response. Some allergic responses are in part due to the breakdown of this sort of regulation.

Antimicrobial agent Any chemical or biological agent that harms the growth of micro-organisms.

Antinutrients Compounds that inhibit normal uptake of nutrients.

Anti-oncogene A gene whose product prevents the normal growth of tissue. *cf* recessive oncogene.

Antioxidant solution Pre-treatment solution (e.g., Vitamin C; citric acid) that retards senescence and browning of tissue. It is employed to incubate explants prior to surface sterilization.

Antioxidant Compound that slows the rate of oxidation reactions.

Apoenzyme Inactive enzyme that has to be associated with a specific organic molecule called a co-enzyme in order to function. The apoenzyme/co-enzyme complex is called a holoenzyme.

Apomixis (Gr. *apo*, away from + *mixis*, a mingling; adj: apomictic) The asexual production of diploid offspring without the fusion of gametes. The embryo develops by mitotic division of the maternal or paternal gamete, or, in the case of plants, by mitotic division of a diploid cell of the ovule.

Apoptosis The process of cell death, which occurs naturally as a part of normal development, maintenance and renewal of tissue in an organism. Apoptosis differs from necrosis, in which cell death is caused by a toxic substance.

Aquaculture Growing of water plants and animals, rather than harvesting them from wherever they happen to grow in rivers or seas. Usually aquaculture uses fresh water; when it uses sea water it can be called mariculture. It is considered to be a part of biotechnology (although peripheral) because it is a new commercial development, and because it often involves growing organisms in large volumes of water, which has similarities to growing large volumes of yeast or bacteria. Biotechnology also provides clean, well-aerated water for the animals to grow in; food, such as krill or powdered synthetic food; and food additives, such as astaxanthins to ensure that fish and prawns have the right colour. Aquaculture has also been used to mass-produce macro- and micro-algae for chemicals, vitamins and pigments. For both animals and plants, biotechnologists have been using genetic methods to produce triploid and tetraploid organisms, and hybrid algae through plant cell fusions. Triploid trout, for example, are sterile, and can be used for biocontrol of weeds without the threat of their being able to breed themselves.

ARS (autonomous replicating sequence) Any eukaryotic DNA sequence that initiates and supports chromosomal replication; they have been isolated in yeast cells. Also called autonomous(ly) replicating segment.

Artificial inembryonation Non-surgical transfer of embryo(s) to a recipient female. As *in vitro* embryo technology develops, artificial inembryonation will gradually replace artificial insemination.

Artificial insemination The deposition of semen, using a syringe, at the mouth of the uterus to make conception possible. It is used in the breeding of domestic animals.

Artificial seed Encapsulated or coated somatic embryos (embryoids) that are planted and treated like seed.

Artificial selection The practice of choosing individuals from a population for reproduction, usually because these individuals possess one or more desirable traits.

Ascus (pl: asci) Reproductive sac in the sexual stage of a type of fungi (*Ascomycetes*) in which ascospores are produced.

Asexual (Gr. *a*, without + L. *sexualis*, sexual) Any type of reproduction not involving meiosis or the union of gametes.

Asexual embryogenesis The sequence of events whereby embryos develop from somatic cells. a.k.a. somatic cell embryogenesis.

Asexual propagation Vegetative, somatic, non-sexual reproduction of a plant without fertilization. *cf* apomixis.

Asexual reproduction Reproduction that does not involve the formation and union of gametes from the different sexes or mating types. It occurs mainly in lower animals, micro-organisms and plants. In plants, asexual reproduction is by vegetative propagation (e.g., bulbs, tubers, corms) and by formation of spores.

Autotrophy Autotrophy is the capacity of an organism to use light as the sole energy source in the synthesis of organic material from inorganic elements or compounds. Autotrophic organisms include green photosynthesizing plants and some photosynthetic bacteria. *cf* heterotrophy.

Auxin (Gr. *auxein*, to increase) A group of plant growth regulators (natural or synthetic) which stimulate cell division, enlargement, apical dominance, root initiation, and flowering. One naturally produced auxin is indole-acetic acid (IAA).

Auxin-cytokinin ratio The relative proportion of auxin to cytokinin present in plant-tissue-culture media. Varying the relative amounts of these two hormone groups in tissue culture formulae affects the proportional growth of shoots and roots *in vitro*. As the ratio is increased (increased auxin or decreased cytokinin), roots are more likely to be produced, and as it is decreased root growth declines and shoot initiation and growth are promoted. This relationship was first recognized by C.O. Miller and F. Skoog in the 1950s.

Auxotroph (Gr. *auxein*, to increase + *trophe*, nourishment) A mutant cell or micro-organism lacking the capacity to form an enzyme or metabolite present in the parental strain, and that consequently will not grow on a minimal medium, but requires the addition of some compound - such as an amino acid or a vitamin - for growth.

Back mutation A second mutation at the same site in a gene as the original mutation. The second mutation restores the wild-type nucleotide sequence.

Bacteriocide; bactericide A chemical or drug that kills bacterial cells.

Bacteriocin A protein produced by bacteria of one strain and active against those of a closely related strain.

Bacteriostat A substance that inhibits or slows down growth and reproduction of bacteria.

Bacterium (Gr. *bakterion*, a stick; pl: bacteria) Common name for the class Schizomycetes: minute (0.5-5mm), unicellular organisms, without a distinct nucleus. Bacteria are prokaryotes, and most of them are identified by means of Gram staining (q.v.). They are classified on the basis of their oxygen requirement (aerobic vs anaerobic) and shape (spherical = coccus; rodlike = bacillus; spiral = spirillum; comma-shaped = vibrio; corkscrew-shaped = spirochaete; filamentous). Bacteria usually reproduce asexually, by simple cell division, although a few undergo a form of sexual reproduction, termed *conjugation*. A few bacteria can photosynthesize (including green-blue cyanobacteria), some are saprophytes and others are parasites and can cause diseases. They are major agents of fermentation, putrefaction and decay, and frequently a source of contamination in tissue culture. In plant pathology, strains of bacteria causing disease in specific plant cultivars are called pathovars (q.v.).

Baculovirus Baculoviruses are a class of insect virus which have been used to make DNA cloning vectors for gene expression in eukaryotic cells. Baculoviruses have a gene which is expressed at very high levels late in their infection cycle, filling the nucleus of the cell with many-sided bodies full of a protein which is not needed to produce more viruses, but is necessary for the virus's spread in the wild. In a vector cloning system, this gene is replaced by one that the biotechnologist wants expressed. Production of the protein can be up to 50% of the cells' protein content, and several proteins can be made simultaneously, so that multi-subunit enzymes can be made by this system. Being an animal expression system, baculoviruses produce proteins that are glycosylated (addition of carbohydrates) like the proteins in animals, making it an attractive option for the production of biopharmaceuticals. In addition, baculoviruses are non-infective and non-pathogenic to vertebrates.

Balanced lethal system A system for maintaining a recessive lethal allele at each of two loci on the same pair of chromosomes. In a closed population with no crossing-over between the loci, only the double heterozygotes for the lethal mutations survive.

Balanced polymorphism Two or more types of individuals maintained in the same breeding population.

Bar A unit used for pressure of fluid. 1 bar = 10^5 Pa. 1 bar is approximately equivalent to 1 atmosphere.

Base A cyclic, nitrogen-containing compound that is one of the essential components of nucleic acids. Exists in five main forms (adenine, A; guanine, G; thymine, T; cytosine, C; uracil, U). A and G have a similar structure and are called purines; T, C and U have a similar structure and are called pyrimidines. A base joined to a ribose sugar joined to a phosphate group is a nucleotide - the building block of nucleic acids.

Base analogues Unnatural purine or pyrimidine bases that differ slightly in structure from the normal bases, but can be incorporated into nucleic acids. They are often mutagenic.

Base collection In PGR: Defined in the International Undertaking on Plant Genetic Resources (FAO, 1983) as a collection of seed stock or vegetative propagating material (ranging from tissue cultures to whole plants) held for long-term security in order to preserve the genetic variation for scientific purposes and as a basis for plant breeding as multiplication and evaluation. *cf* active collection.

Base pair (bp) The two strands that constitute DNA are held together by specific hydrogen bonding between purines and pyrimidines (A pairs with T; and G pairs with C). The size of a nucleic acid molecule is often described in terms of the number of base pairs (symbol: bp) or thousand base pairs (kilobase pairs; symbol: kb; a more convenient unit) it contains.

Bio-assay A procedure for the assessment of a substance by measuring its effect in living cells or on organisms. Animals have been used extensively in drug research in bio-assays for the pharmacological activity of drugs. However, bio-assays are now usually developed using bacteria or animal or plant cells, as these are usually much easier to handle than whole animals or plants, are cheaper to make and keep, and avoid the ethical problems associated with testing of animals. Sometimes used to detect minute amounts of substances that influence or are essential to growth.

Bio-augmentation Increasing the activity of bacteria that decompose pollutants; a technique used in bioremediation.

Bio-energetics The study of the flow and the transformations of energy that occur in living organisms.

Bio-engineering The use of artificial tissues, organs and organ components to replace parts of the body that are damaged, lost or malfunctioning.

Bio-enrichment Adding nutrients or oxygen to increase microbial breakdown of pollutants.

Bio-ethics The branch of ethics that deals with the life sciences and their potential impact on society. At one extreme, it can be enormously useful in focusing attention on problems that need to be confronted; at the other extreme, it can become a name-calling argument between the "pro-biotechnology" and "anti-biotechnology" schools of thought, which, as it reduces discussion to epithets and clichés, can make better sound bites.

Biofuel A gaseous, liquid or solid fuel that contains energy derived from a biological source. For example, rapeseed oil or fish liver oil can be used in place of diesel fuel in modified engines. A commercial application is the use of modified rapeseed oil, which - as rapeseed methyl ester (RME) - can be used in modified diesel engines, and is sometimes named bio-diesel. *cf* biogas.

Biogas A mixture of methane and carbon dioxide resulting from the anaerobic decomposition of waste such as domestic, industrial and agricultural sewage. a.k.a. *gobar.*

Biogenesis The principle that a living organism can only arise from other living organisms similar to itself and can never originate from non-living material.

Bio-informatics The use and organization of information of biological interest. In particular, it is concerned with organizing bio-molecular databases, in getting useful information out of such databases, in utilizing powerful computers for analysing such information, and in integrating information from disparate biological sources.

Biolistics (from **bio**logical + bal**listics**) A technique to insert DNA into cells. The DNA is mixed with small metal particles - usually tungsten or gold - a fraction of a micrometre across. These are then fired into a cell at very high speed. They puncture the cell and carry the DNA into the cell. Biolistics has an advantage over transfection, transduction, etc., because it can apply to any cell, or indeed to parts of a cell. Thus use of biolistics has inserted DNA into animal, plant and fungal cells, and into mitochondria inside cells. a.k.a. microprojectile bombardment.

Biological containment Restricting the movement of (genetically engineered) organisms by arranging barriers to prevent them from growing outside the laboratory. Biological containment can take two forms: making the organism unable to survive in the outside environment, or making the outside environment inhospitable to the organism. The latter is rarely suitable for bacteria, which, in principle, could survive almost anywhere. Thus for bacteria and yeasts, the favoured approach is to mutate the genes in the organism so that they require a supply of a specific nutrient that is usually available only in the laboratory. If they get out, they then cannot grow. Making the environment unfriendly to the organism is partly a biological control, partly a physical one. Thus, some of the first genetically engineered rice strains were developed in England (which is too cold for rice to grow) and tried in the field in Arizona (where it is too dry). Biological containment may also involve the use of vector molecules and host organisms which have been genetically disabled such that they can survive only in the peculiar conditions provided by the experimenter and which are unavailable outside the laboratory.

Biome A major ecological community or complex of communities, extending over a large geographical area and characterized by a dominant type of vegetation.

Biometry The application of statistical methods to the analysis of biological problems.

Biopesticide A compound that kills organisms by virtue of specific biological effects rather than as a broader chemical poison. Specific types include bio-insecticides and bio-fungicides. Bio-pesticides differ from biocontrol agents in that bio-pesticides are passive agents, whereas biocontrol agents are active, seeking out the pest to be destroyed. There are some extremely attractive anti-pest materials, such as *Bacillus thuringiensis* (Bt) toxin, which specifically interferes with the absorption of food from the guts of some insects but is harmless to mammals. The rationale behind developing bio-pesticides is that they are more likely to be biodegradable and are targeted at specific elements of the pest's metabolism.

Biopolymer Any large polymeric molecule (protein, nucleic acid, polysaccharide, lipid) produced by a living organism.

Bioprocess Any process that uses complete living cells or their components (e.g., enzymes, chloroplasts) to effect desired physical or chemical changes.

Bioreactor A tank in which cells, cell extracts or enzymes carry out a biological reaction. Often refers to a growth chamber (fermenter, fermentation vessel) for cells or micro-organisms.

Bioremediation A process that uses living organisms to remove contaminants, pollutants or unwanted substances from soil or water. *cf* bio-augmentation; bio-enrichment.

Biosensor A device that uses an immobilized agent (such as an enzyme, antibiotic, organelle or whole cell) to detect or measure a chemical compound. A reaction between the immobilized agent and the molecule being analysed is transduced into an electric signal.

Biosphere The part of the earth and its atmosphere that is inhabited by living organisms.

Biosynthesis Synthesis of compounds by living cells, which is the essential feature of anabolism.

Biotechnology The use of biological processes or organisms for the production of materials and services of benefit to humankind. Biotechnology includes the use of techniques for the improvement of the characteristics of economically important plants and animals and for the development of micro-organisms to act on the environment.

Biotic factor Other living organisms that are a factor of an organism's environment, and form the biotic environment, affecting the organism in many ways.

Biotic stress Stress resulting from living organisms which can harm plants, such as viruses, fungi, bacteria, parasitic weeds and harmful insects. *cf* abiotic stress.

Biotin A vitamin of the B complex. It is a co-enzyme for various enzymes that catalyse the incorporation of carbon dioxide into various compounds. It is essential for the metabolism of fats. Biotin is attached to pyruvate carboxylase by a long, flexible chain like that of lipoamide in the pyruvate dehydrogenase complex. Adequate amounts are normally produced by the intestinal bacteria in animals. a.k.a. vitamin H (in USA).

Biotinylated-DNA A DNA molecule labelled with biotin by incorporation of biotinylated-dUTP into a DNA molecule. It is used as a non-radioactive probe in hybridization experiments, such as Southern transfer. The detection of the labelled DNA is achieved by complexing it with streptavidin (an antibiotic with a high affinity for biotin) to which is attached a colour-generating agent such as horseradish peroxidase that gives a fluorescent green colour upon reaction with various organic reagents.

Biotransformation The conversion of one chemical or material into another using a biological catalyst: a near synonym is biocatalysis, and hence the catalyst used is called a biocatalyst. Usually the catalyst is an enzyme, or a whole, dead micro-organism that contains an enzyme or several enzymes.

Biotope A small habitat in a large community.

Biotoxin A naturally produced toxic compound which shows pronounced biological activity and presumably has some adaptive significance to the organism which produces it.

Bivalent A pair of synapsed or associated homologous chromosomes (one of maternal origin; the other of paternal origin) that have each undergone duplication. Each duplicated chromosome comprises two chromatids. Thus a bivalent comprises four chromatids.

Blastomere Any one of the cells formed from the first few cleavages in animal embryology. The embryo usually divides into two, then four, then eight blastomeres, and so on.

Blastula In animals, an early embryo form that follows the morula stage; typically, a single-layered sheet (blastoderm) or ball of cells (blastocyst).

B lymphocytes; B cells An important class of lymphocytes that mature in bone marrow (in mammals) and the Bursa of Fabricius (in birds), that are largely responsible for the antibody-mediated or humoral immune response; they give rise to the antibody-producing plasma cells and some other cells of the immune system.

Bleach A fluid, powder or other whitening (bleaching) or cleaning agent, usually with free chlorine ions. Commercial bleach contains calcium hypochlorite or sodium hypochlorite, and is a common disinfectant used for cleaning working surfaces, tools and plant materials in plant tissue culture and grafting.

Bleeding Used to describe the occasional purplish-black coloration of media due to phenolic products given off by (usually fresh) transfers.

Blunt end The end of a DNA duplex molecule in which neither strand extends beyond the other. a.k.a. flush end.

Blunt-end cut To cleave phospho-diester bonds in the backbone of duplex DNA between the corresponding nucleotide pairs on opposite strands. This cleavage process results in both strands finishing at the same residue, i.e., there are no nucleotide extensions on either strand. a.k.a. flush-end cut.

Blunt-end ligation Joining (ligation) of the nucleotides that are at the ends of two blunt-ended DNA duplex molecules.

Breed Noun: In AnGR, either (i) a sub-specific group of domestic livestock with definable and identifiable external characteristics that enable it to be separated by visual appraisal from other similarly defined groups within the same species or (ii) a group of domestic livestock for which geographical and/or cultural separation from phenotypically similar groups has led to acceptance of its separate identity. *cf* breed at risk; breed not at risk; critical breed; critical-maintained breed; endangered-maintained breed. (Source: FAO, 1999)

Browning Discoloration due to phenolic oxidation of freshly cut surfaces of explant tissue. In later stages of culture, such discoloration may indicate a nutritional or pathogenic problem, generally leading to necrosis.

Brucellosis Disease caused by infection with organisms of the genus *Brucella*.

Bubble column fermenter A fermentation vessel, or bioreactor, in which the cells or micro-organisms are kept suspended in a tall cylinder by rising air, which is introduced at the base of the vessel.

Bud A region of meristematic tissue with the potential for developing into leaves, shoots, flowers or combinations; generally protected by modified scale leaves. A terminal (or apical) bud exists at the tip of a stem or branch, while axillary (or lateral) buds develop in the axils of leaves.

Bud scar A scar left on a shoot when the bud or bud scales drop.

Bud sport A somatic mutation arising in a bud and producing a genetically different shoot. Bud sports includes changes due to gene mutation, somatic reduction, chromosome deletion or polyploidy.

Buffer A solution that resists change in pH when an acid or alkali is added, or when solutions are diluted.

Calf scours. A watery diarrhoea in calves.

Calorie (abbr: cal) Equivalent to the amount of heat required to raise the temperature of 1 gram of water from 14.5°C to 15.5°C (»4.19 J). *cf* kilocalorie.

Calyx (Gr. *kalyx*, a husk, cup) All the sepals of a flower considered collectively. The outermost whorl of flower parts.

Cambial zone Region in stems and roots consisting of the cambium and its recent derivatives.

Cambium (L. *cambium*, one of the alimentary body fluids supposed to nourish the body organs; pl: cambia). A layer, usually regarded as one or two cells thick, of persistently meristematic tissue between the xylem and phloem tissues, and which gives rise to secondary tissues, thus resulting in an increase in diameter. The two most important cambia are the vascular (fascicular) cambium and the cork cambium.

Cancer Uncontrolled growth of the cells of a tissue or an organ in a multicellular organism. *cf* oncogenesis.

Candidate gene A gene whose function suggests that it may be involved in the genetic variation observed for a particular trait, e.g., the gene for growth hormone is a candidate gene for body weight.

Candidate-gene strategy An experimental approach in which knowledge of the biochemistry and/or physiology of a trait is used to draw up a list of genes whose protein products could be involved in the trait.

Canine Pertaining to dogs.

Canola Any of several cultivars of oilseed rape (more fully: canola oil); the vegetable oil high in mono-unsaturated fatty acid obtained from these cultivars.

Cap site The site in a gene where translation is initiated, a.k.a. translation initiation site.

Capacitation The final stage in the maturation process of a spermatozoon, taking place inside the female genital tract as the sperm penetrates the ovum.

Carbohydrate An organic compound based on the general formula $C_x (H_2O)_y$, performing many vital roles in living organisms. The simplest carbohydrates are the sugars (saccharides), including glucose and sucrose. Polysaccharides are carbohydrates of much greater molecular weight and complexity; examples are starch, which serves as energy store in plant seeds and tubers; cellulose and lignin that form the cell walls and woody tissue of plants of plants; glycogen, etc.

Carboxypeptidases Two enzymes (A and B) found in pancreatic juice. Their role is to remove the C-terminal amino-acid from a peptide; the A form removes any amino acid; the B form removes only lysine or arginine. Used when sequencing peptides.

Carpel Female reproductive organ of flowering plants, consisting of stigma, style and ovary. In some plants, one or more carpels unite to form the pistil.

Carrier In genetics, typically an individual that has one recessive mutant allele for some defective condition that is "masked" by a dominant normal allele at the same locus, i.e., an individual that is heterozygous for a recessive harmful allele and a dominant normal allele; the phenotype is normal, but the individual passes the defective (recessive) allele to half of its offspring.

Carrier gas The gas that carries the sample in gas chromatography.

Catabolite repression Glucose-mediated reduction in the rates of transcription of operons that encode enzymes involved in catabolic pathways (such as the *lac* operon).

Catalyst (Gr. *katalyein*, to dissolve) A substance that promotes a chemical reaction by lowering the activation energy of a chemical reaction, without itself undergoing any permanent chemical change. The process is catalysis. *cf* catalytic antibody; catalytic RNA.

Catalytic antibody (= abzyme) An antibody selected for its ability to catalyse a chemical reaction by binding to and stabilizing the transition-state intermediate.

Catalytic RNA (= ribozyme; gene shears) A natural or synthetic RNA molecule that cuts an RNA substrate.

Cation A positively charged ion; opposite is *anion*.

Caulogenesis Stem organogenesis; induction of shoot development from callus.

CD molecules; cluster of differentiation molecules Any group of antigens that is associated with a specific sub-population of T cells. There are designations for surface molecules on various cells of the immune system, e.g., CD4 is present on the surface of helper T cells.

CDNA clone A double-stranded DNA molecule that is carried in a vector and was synthesized *in vitro* from an mRNA sequence by using reverse transcriptase and DNA polymerase.

CDNA cloning A method of cloning the coding sequence of a gene, starting with its mRNA transcript. It is normally used to clone a DNA copy of a eukaryotic mRNA. The cDNA copy, being a copy of a mature messenger molecule, will not contain any intron sequences and may be readily expressed in any host organism if attached to a suitable promoter sequence within the cloning vector.

CDNA library A collection of cDNA clones that were generated *in vitro* from the mRNA sequences isolated from an organism or a specific tissue or cell type or population of an organism. *cf* library .

CDR (complementarity-determining regions) These are regions of the variable (V) regions of light and heavy antibody chains that make contact with the antigen. The primary amino acid sequences of these regions are highly variable among antibodies of the same class.

Cell (L. *cella*, small room) The smallest structural unit of living matter capable of functioning independently; a microscopic mass of protoplasm surrounded by a semi-permeable membrane, usually including one or more nuclei and various non-living products, capable - either alone or by interacting with other cells - of performing all the fundamental functions of life.

Cell culture The *in vitro* growth of cells derived from multi-cellular organisms. The cells are usually of one type.

Cell cycle Sequence of stages that a cell passes through between one division and the next. The cell cycle oscillates between mitosis and the interphase, which is divided into G, S, and G_2. In the G phase there is a high rate of biosynthesis and growth; in the S phase there is the doubling of the DNA content as a consequence of chromosome replication; in the G_2 phase the final preparations for cell division (cytokinesis) are made.

Cell differentiation Continuous loss of physiological and cytological characters of young cells, resulting in getting the characters of adult cells. The unspecialized cells become modified and specialized for the performance of specific functions. Differentiation results from the controlled activation and de-activation of genes.

Cell division Formation of two or more daughter cells from a single mother cell. The nucleus divides first, followed by the formation of a cell membrane between the daughter nuclei. Division of cytoplasm and nucleus into two or more parts by formation of a cell plate.

Cell fusion Formation of a single hybrid cell from two cells of different species, cultured *in vitro*. The cells fuse and coalesce, but their nuclei may remain separated. During subsequent cell division, a single spindle is formed so that each daughter cell has a single nucleus containing sets of chromosomes from each parental line. Subsequent divisions often result in the loss of chromosomes and therefore of genes. The cell fusion technique can be used to determine the control of specific genes and their assignment to chromosomes. *cf* cell hybridization.

Cell generation time The interval between the beginning of consecutive divisions of a cell. The time that it takes for a population of single-celled organisms to double its cell number. Successive generations of cells or organisms within a population are separated by a time interval called generation time. The cell regeneration time can be determined with the aid of time-lapse microcinematography.

Cell hybridization The fusion of two or more dissimilar cells leading to the formation of a somatic hybrid. *cf* cell fusion.

Cell line A cell lineage that can be maintained in culture. A cell line arises from a primary culture. It implies that cultures from it consist of several lineages of the cells originally present in the primary culture.

Cell-mediated immune response The activation of T cells of the immune system in response to the presence of a foreign antigen.

Cell number The number of cells per unit volume of a culture.

Cell plate The precursor of the cell wall, formed as cytokinesis starts during cell division. The cell plate develops in the region of the equatorial plate and arises from membranes in the cytoplasm.

Chemostat A continuous and open culture in which growth rate and cell density are maintained constant by a fixed rate of input of a growth-limiting nutrient. *cf* phytostat.

Chemotaxis Motion of a motile cell, organism or part towards or away from an increasing concentration of a particular substance.

Chemotherapy The treatment of disease, especially infections or cancer, by means of chemicals. In treating cancers, it involves administering chemicals toxic to malignant cells.

Chiasma (Gr. *chiasma*, two lines placed crosswise; pl. chiasmata) A visible point of junction between two non-sister chromatids of homologous chromosomes during the first meiotic prophase. a.k.a. cross-over. In the diplotene stage of prophase I of meiosis, the four chromatids of a bivalent are associated in pairs, but in such a way that one part of two chromatids is exchanged.

Chimeric selectable marker gene A gene that is constructed from parts of two or more different genes and allows the host cell to survive under conditions where it would otherwise die.

Chi-squared test (c^2 test) A significance test used to statistically assess the goodness of fit of observed data to a prediction.

Chitin A nitrogenous polysaccharide occurring as skeletal material in many invertebrates and fungi.

Chitinase An enzyme which breaks down chitin.

Chloramphenicol An antibiotic that interferes with protein synthesis.

Chlorenchyma (Gr. *chloros*, green + *enchyma*, a suffix meaning tissue) Tissue containing chloroplasts, including leaf mesophyll and other parenchyma cells.

Chlorophyll (Gr. *chloros*, green + *phyllon*, leaf) One of the two pigments responsible for the green colour of most plants. It is essential in the absorption of light energy for photosynthesis.

Chloroplast (Gr. *chloros*, green + *plastos*, formed) Specialized cytoplasmic organelle that contains chlorophyll. Lens-shaped and bounded by a double membrane, chloroplasts contain membranous structures (thylakoids) piled up into stacks, surrounded by a gel-like matrix (stroma). They are the site of solar energy transfer and important reactions of starch or sugar synthesis. Chloroplasts have their own DNA and are inherited cytoplasmically, independent of nuclear genes.

Chlorosis (Gr. *chloros*, green + *osis*, diseased state) Failure of chlorophyll development, and appearance of yellow colour in plants, because of a nutritional disturbance or because of an infection by a virus, bacteria or fungus.

Chromatid (chromosome + *id*, L. suffix meaning "daughters of") Each of the two daughter strands comprising a duplicated chromosome. The term remains in use while the two chromatids are still joined at the centromere. As soon as the centromere divides, setting the two chromatids adrift (during anaphase of mitosis; and during anaphase II of meiosis), they are called chromosomes.

Chromatin (Gr. *chroma*, colour) Substance of which eukaryotic chromosomes are composed. It consists of primarily DNAm with some proteins (mainly histones), and small amounts of RNA. Originally named because of the readiness with which it stains with certain dyes (chromaticity).

Chromatin fibres A basic organizational unit of eukaryotic chromosomes, consisting of DNA and associated proteins assembled into strands of 30 nm average diameter.

Chromomeres Small bodies, described by J. Belling, that are identified by their characteristic size and linear arrangement along a chromosome.

Chromonema (pl: chromonemata) An optically single thread forming an axial structure within each chromosome.

Chromoplast Plastid containing pigments, such a chloroplast, or one in which carotenoids predominate.

Chromosomal aberration Any change in chromosome structure or number. Although it can be a mechanism for enhancing genetic diversity, such alterations are usually fatal or ill-adaptive, especially in animals.

Chromosomal integration site A chromosomal location where foreign DNA can be integrated, often without impairing any essential function in the host organism.

Chromosomal polymorphism The occurrence of one to several chromosomes in two or more alternative structural forms within a population; the structurally changed chromosomes are the result of chromosome mutations (i.e., any structural change involving the gain, loss or re-location of chromosome segments).

Codon optimization An experimental strategy in which codons within a cloned gene - ones not generally used by the host cell translation system - are changed by *in vitro* mutagenesis to the preferred codons, without changing the amino acids of the synthesized protein.

Coefficient A number expressing the amount of some change or effect under certain conditions (e.g., the coefficient of inbreeding).

Co-enzyme An organic molecule of low molecular weight and usually non-protein, such as a vitamin, that binds to an enzyme and promotes its catalytic activity.

Co-evolution The evolution of complementary adaptations in two species caused by the selection pressure that each exerts on the other. It is common in symbiotic associations, in insect-pollinated plants, etc.

Co-factor An organic molecule or inorganic ion necessary for the normal catalytic activity of an enzyme.

Co-fermentation The simultaneous growth of two micro-organisms in one bioreaction.

Co-generation Production of both electricity and process heat (steam) in an industrial plant.

Cohesion Holding together; a force holding a solid to a solid or a solid to a liquid, owing to attraction between like molecules.

Coincidence The ratio of the observed frequency of double crossings-over to the expected frequency, where the expected frequency is calculated by assuming that the two crossing-over events occur independently of each other.

Co-integrate vector system A two-plasmid system for transferring cloned genes to plant cells. The cloning vector has a T-DNA segment that contains cloned genes. After introduction into *Agrobacterium tumiefasciens*, the cloning vector DNA undergoes homologous recombination with a resident disarmed Ti plasmid to form a single plasmid carrying the genetic information for transferring the genetically engineered T-DNA region to plant cells.

Co-integrate A DNA molecule formed by the fusion of two different DNA molecules, usually mediated by a transposable element.

Colchicine (L. *colchicum*, meadow saffron, from colchis, ancient Mingrelia). An alkaloid obtained from *Colchicum autumnale*, autumn crocus or meadow saffron, which inhibits spindle formation in cells during mitosis, so that chromosomes cannot separate during anaphase, thus inducing multiple sets of chromosomes. Also used to halt mitosis at metaphase - the stage when chromosomes are most visible.

Coleoptile Protective sheath covering the shoot apex of the embryo in monocotyledenous plants. It protects the plumule as it emerges through the soil.

Coleorhiza (Gr. *koleos*, sheath + *rhiza*, root) A protective sheath surrounding the radicle of monocotyledenous plants.

Co-linearity A relationship in which the units in one molecule occur in the same sequence as the units in another molecule which they specify; e.g., the nucleotides in a gene are co-linear with the amino acids in the polypeptide encoded by that gene.

Collection In PGR: *see* base collection; active collection.

Collenchyma (Gr. *kolla*, glue + *enchyma*, a suffix, derived from parenchyma and denoting a type of cell tissue) A tissue of living cells, the walls being unevenly thickened with cellulose and hemicellulose, but never lignified; it functions in mechanical support in young, short-lived or non-woody organs and is thus found in midribs and leaf petioles.

Companion cell Living cell associated with the sieve cell of phloem tissue in vascular plants.

Comparative gene mapping The comparison of map locations of genes between species. The results of these comparisons indicate substantial conservation of blocks of genes and even large segments of chromosomes between species. Great use can be made of this conservation of map position. For example, in the case of mammals, it means that if a gene has been mapped in one or both of the intensely-mapped species (humans and mice), then the likely location of that gene in other mammals can be predicted with considerable confidence. Conversely, if a mapped anonymous DNA marker has an effect on a quantitative trait (this being indicative of the marker being linked to a quantitative trait locus (QTL)) in, say, cattle, then knowledge of the comparative map between cattle and humans can identify genes in the homologous region of the human genome that could correspond to the QTL. Such genes are called comparative positional candidate genes (q.v.).

Comparative positional candidate gene A gene that is likely to be located in the same region as a DNA marker that has been shown to be linked to a single-locus trait or to a quantitative trait locus (QTL), where the gene's likely location in the genome of the species in question is based on its known location in the map of another species, i.e., is based on the comparative map between the two species.

Competence Ability of a bacterial cell to take up DNA molecules and become genetically transformed.

Competency An ephemeral state, induced by treatment with cold cations, during which bacterial cells are capable of taking up foreign DNA.

Competent A competent cell is capable of developing into a fully functional embryo. The opposite is non-competent.

Complement proteins Proteins that bind to antibody-antigen complexes and help degrade the complexes by proteolysis.

Complementarity The relationship between the two strands of a double helix of DNA. Thymine in one strand pairs with adenine in the other strand, and cytosine in one strand pairs with guanine in the other strand.

Complementation test; trans test Introduction of two mutant chromosomes into the same cell to determine whether the mutants are alleles of the same gene. If the mutations are non-allelic, the genotype will be m_1 +/+ M_2, and the phenotype will be wild-type (normal), because each chromosome "covers" for the other. In contrast, if they are allelic, the mutant phenotype will result.

Complete digest The treatment of a DNA preparation with an endonuclease for sufficient time for all of the potential target sites within that DNA to have been cleaved. *cf* partial digest.

Composite transposon A transposable element formed when two identical or nearly identical transposons insert on either side of a non-transposable segment of DNA, such as the bacterial transposon Tn5.

Compound chromosome A chromosome formed by the union of two separate chromosomes, as in attached-X chromosomes or attached-X-Y chromosomes.

Concatemer A DNA segment made up of repeated sequences linked end to end.

Contig A set of overlapping clones that provide a physical map of a portion of a chromosome. *cf* contiguous map.

Continuous culture A suspension culture continuously supplied with nutrients by the inflow of fresh medium. The culture volume is normally constant. *cf* batch culture; closed continuous culture; continuous fermentation; open continuous culture.

Continuous fermentation A process in which cells or micro-organisms are maintained in culture in the exponential growth phase by the continuous addition of fresh medium that is exactly balanced by the removal of cell suspension from the bioreactor. *cf* continuous culture.

Continuous variation Variation not represented by distinct classes. Phenotypes grade into each other, and measurement data are required for analysis. Multiple genes are usually responsible for this type of variation. a.k.a. quantitative variation. *cf* discontinuous variation.

Coordinate repression Correlated regulation of the structural genes in an operon by a molecule that interacts with the operator sequence.

Co-polymers Mixtures consisting of more than one monomer; for example, polymers of two kinds of organic bases, such as uracil and cytosine (poly-UC) have been combined for studies of the genetic code.

Copy number The average number of molecules of a plasmid or gene per genome contained in a cell.

Co-repressor An effector molecule that forms a complex with a repressor and turns off the expression of a gene or set of genes.

Correlation A statistical association between variables.

Cortex Primary tissue of a stem or root, bounded externally by the epidermis and internally in the stem by the phloem and in the root by the pericycle.

Cos ends The 12-base, single-strand, complementary extensions of bacteriophage lambda (l) DNA. a.k.a. cos sites.

Co-segregation When two genetic conditions appear to be inherited together.

Cosmid A plasmid vector which contains the two cos (cohesive) ends of phage lambda (l) and one or more selectable markers such as an antibiotic resistance gene. Cosmids exploit certain properties of phage lambda (l) to enable large, 40-50 kb, DNA fragments to be cloned at high efficiency. Cosmids and cosmid recombinants replicate as plasmids.

Cot curve When duplex DNA is heated, it dissociates into single strands. When the temperature is lowered, complementary strands tend to anneal or re-nature. The extent of re-naturation depends on the product of DNA concentration in moles of nucleotides per litre, and time in seconds. A graph showing the proportion of re-natured DNA against cot is known as a cot curve. The cot at which half the DNA has re-natured is the half-cot, a parameter indicating the degree of complexity of the DNA.

Co-transfection In baculovirus expression systems, the procedure by which the baculovirus and the transfer vector are simultaneously introduced into insect cells in culture.

Co-transformation In genetic engineering experiments, it is often necessary to transform with a plasmid for which there is no selectable phenotype and then screen for the presence of that plasmid within the host cell. Co-transformation is a technique in which host cells are incubated with two types of plasmid, one of which is selectable and the other not. Cells which have been transformed with the first plasmid are then selected. If transformation has been carried out at high DNA concentration, then it is probable that these cells will also have been transformed with the second (non-selectable) plasmid. The technique is frequently used in experiments with mammalian cells.

Cotyledons Leaflike structures at the first node of the seedling stem. In some dicotyledons, they contain stored food for the young plant not yet able to photosynthesize its own food. Often referred to as seed leaves.

Coupling The phase state in which either two dominant or two recessive alleles of two different genes occur on the same chromosome. Also called cis configuration. *cf* repulsion.

Covalent bond A bond in which an electron pair is equally shared by protons in two adjacent atoms.

Covalently closed circle (CCC) A double-stranded DNA molecule with no free ends. The two strands are interlinked and will remain together even after denaturation. In its native form, a CCC will adopt a supercoiled configuration.

Co-variance A measure of the statistical association between variables; the extent to which two variables vary together.

CpDNA The DNA of plant plastids, including chloroplasts.

Critical breed In AnGR: A breed where the total number of breeding females is less than 100 or the total number of breeding males is less than or equal to five; or the overall population size is close to, but slightly above 100 and decreasing, and the percentage of pure-bred females is below 80%. (Source: FAO, 1999)

Critical-maintained breed; endangered-maintained breed In AnGR: Categories where critical or endangered breeds are being maintained by an active public conservation programme or within a commercial or research facility. (Source: FAO, 1999)

Cross In genetic studies, the mating of two individuals or populations. Also called mating.

Cross-breeding Mating between members of different populations (lines, breeds, races or species).

Cross hybridization The hydrogen bonding of a single-stranded DNA sequence that is partially but not entirely complementary to a single-stranded substrate. Often, this involves hybridizing a DNA probe for a specific DNA sequence to the homologous sequences of different species.

Crossing over A process in which homologous chromosomes exchange material through the breakage and reunion of two chromatids. A single crossover represents one reciprocal breakage and reunion event. A double crossover requires two simultaneous reciprocal breakage and reunion events. a.k.a. recombination; recombination event.

Crossover A recombinant chromosome.

Crossing-over unit A measure of distance between two loci on genetic maps that is based on the average number of crossing-over events that take place in the interval between those two loci during meiosis. A map interval that is one crossing-over unit in length (a centiMorgan) describes an interval between two loci such that one in every hundred gametes recovered from meiosis is recombinant in that interval, i.e., the allele at the first locus is maternal in origin, while the allele at the other locus in that same gamete is paternal in origin.

Cross pollination Fertilization of a plant from a plant with a different genetic makeup.

Cross pollination efficiency Efficiency of pollen from one plant reaching the stigma of another plant.

Crown The region at the base of the stem of cereals and forage species from which tillers or branches arise. In woody plants, the root-stem junction. In forestry, the top portions of the tree.

Cryogenic At very low temperature.

Cryoprotectant Compound preventing cell damage during freezing and thawing processes. Cryoprotectants are agents with high water solubility and low toxicity. Two types of cryoprotectant agent are commonly used: permeating (glycerol and DMSO (q.v.)) and non-permeating (sugars, dextran, ethylene glycol, polyvinyl pyrolidone and hydroxyethyl starch).

Culture A population of plant or animal cells or micro-organisms that is grown under controlled conditions.

Culture alteration A term used to indicate a persistent change in the properties of a culture's behaviour (e.g., altered morphology, chromosome constitution, virus susceptibility, nutritional requirements, proliferative capacity, etc.). The term should always be qualified by a precise description of the change which has occurred in the culture.

Culture medium Any nutrient system for the cultivation of cells of plants, bacteria or other organisms; usually a complex mixture of organic and inorganic nutrients. *cf* medium.

Culture room Room for maintaining cultures and often in a controlled environment of light, temperature and humidity. *cf* growth cabinet; incubator.

Curing The elimination of a plasmid from its host cell. Many agents which interfere with DNA replication, e.g., ethidium bromide, can cure plasmids from either bacterial or eukaryotic cells.

Cut Slang: to make a double-stranded break in DNA, usually with a type II restriction endonuclease. E.g., "The DNA was cut with *Eco*RI and run out on a 1% agarose gel." *cf* nick; cleave.

Cytoplasmic genes DNA-containing bodies in the cell but external to the nucleus.

Dabs (single-domain antibodies) Antibodies with only one (instead of two) protein chain derived from only one of the domains of the antibody structure. Dabs exploit the finding that, for some antibodies, half of the antibody molecule binds to its target antigen almost as well as the whole molecule. The potential advantages of dabs are that they can be made easily by bacteria or yeasts, and offer a way to clone antibody-like molecules into bacteria, and hence to be able to easily screen millions of antibodies. Related ideas are single-chain antigen binding technology (SCA), biosynthetic antibody binding sites (BABS), minimum recognition units (MRUs), and complementary determining regions (CDRs).

Dalton (symbol: Da) A unit of atomic mass roughly equivalent to the mass of a hydrogen atom. 1.67×10^{-24} g. Named after the famous nineteenth-century chemist, John Dalton (1766-1844)). Used in shorthand expressions of molecular weight, especially as kilo- (kDa) or megadaltons (MDa), which are equal to respectively to 1×10^3 and 1×10^6 daltons.

Deficiency Insufficiency or absence of one or more usable forms of enzymatic, nutritional or environmental requirements, so that development, growth or physiological functions are affected.

Dehiscence The spontaneous and often violent opening of a fruit, seed pod or anther to release and disperse the seeds or pollen.

Dehydrogenase An enzyme that catalyses the remove of hydrogen atoms in biological reactions.

Dehydrogenation A chemical reaction in which hydrogen is removed from a compound.

De-ionized water Water which is free of most inorganic (not completely free, since Na is present in ample quantities) and most organic compounds.

Deletion A mutation involving the removal of one or more base pairs in DNA sequence. Large deletions are visible as the lack of chromosomal segments.

Deliberate release Putting something into the outside world; in biotechnology it means putting a genetically modified organism (GMO) into field trials.

Deme A group of organisms in the same taxon.

De-mineralize To remove the mineral content (salts, ions) from a substance, especially water. Removal methods include distillation and electrodialysis. The process is de-mineralization.

Denaturated DNA Duplex DNA that has been converted to single strands by breaking the hydrogen bonds of complementary nucleotide pairs. Usually achieved by heating.

Denaturation Loss of native configuration of a macro-molecule (protein or nucleic acid) by physical or chemical means, usually accompanied by loss of biological activity. Denatured proteins often unfold their polypeptide chains and express changed properties of solubility. The separation of duplex nucleic acid molecules into single strands. Most commonly used by genetic engineers to describe the destruction of hydrogen bonds maintaining the double-stranded nature of all or part of a DNA molecule.

Desulphurization (USA: desulfurization) Technology for removing sulphur from oil and coal by use of bacteria. Sulphur residues in fuels end up as sulphur dioxide when the fuel is burned, resulting in acid rain. Bacteria may oxidize sulphites (insoluble) into sulphates (soluble), which can be washed away with the bacteria.

Determinate growth Growth determined and limited in time, as in most floral meristems and leaves. The differentiation process is irreversibly established. Determinate growth contrasts with the usual culture growth, which is infinite and indeterminate.

Determination Process by which undifferentiated cells in an embryo become committed to develop into specific cell types, such as neurons, fibroblasts or muscle cells.

Determined Describing embryonic tissue at a stage when it can develop only as a certain kind of tissue.

Development The sum total of events that contribute to the progressive elaboration of an organism. The two major aspects of development are growth and differentiation.

Diagnostic procedure A test or assay used to determine the presence of an organism, substance or nucleic acid sequence alteration.

Diakinesis A stage of meiosis just before metaphase I, in which the separation of homologous chromosomes is almost completed.

Diazotroph An organism that can fix atmospheric nitrogen.

Dicentric chromosome A chromosome having two centromeres.

Dichogamy The condition in which the male and the female reproductive organs of a flower mature at different times, thereby making self-fertilization improbable or impossible.

Dicotyledon (Gr. *dis*, twice + *kotyledon*, a cup-shaped hollow) A plant with two cotyledons, or seed leaves. One of the two classes of plants in the Angiosperms (the other class is the monocotyledons). Colloquially called a dicot. Examples include many crop plants (potato, pea, beans), ornamentals (rose, ivy) and timber trees (oak, beech, lime).

Differential centrifugation A method of separating sub-cellular particles according to their sedimentation coefficients, which are roughly proportional to their size. Cell extracts are subjected to a succession of centrifuge runs at progressively faster rotation speeds. Large particles, such as nuclei or mitochondria, will be precipitated at relatively slow speeds; higher G forces will be required to sediment small particles, such as ribosomes.

Differentially permeable Referring to a membrane, through which different substances diffuse at different rates; some substances may be unable to diffuse through such a membrane.

Differentiation (L. *differre*, to carry different ways) A process in which unspecialized cells develop structures and functions characteristic of a particular type of cell. Development from one cell to many cells, accompanied by a modification of the new cells for the performance of particular functions. In tissue culture, the term is used to describe the formation of different cell types.

Discordant Members of a pair showing different, rather than similar, characteristics.

Disease (L. *dis*, a prefix signifying the opposite + M.E. *aise*, comfort) The opposite of ease. Any alteration from the state of metabolism necessary for the normal development and functioning of an organism, usually associated with infection by a pathogen or the malfunction or absence of one or more genes.

Disease resistance The ability to remain healthy by resisting disease or the disease agent. Disease resistance or tolerance is a subject of intense interest in biotechnology.

Disease-free A plant or animal certified through specific tests as being free of specified pathogens. Disease-free should be interpreted to mean "free from any known diseases" as "new" diseases may yet be discovered to be present. *cf* disease indexing.

Disease-indexing Disease-indexed organisms have been assayed for the presence of known diseases according to standard testing procedures. *cf* disease-free.

Disinfectation Full elimination of internal micro-organisms from a culture; disinfectation is rarely obtained. *cf* disinfestation.

Disinfestation The elimination or inhibition of the activity of surface-adhering micro-organisms. *cf* disinfection.

Disomy (adj: disomic) The presence of a pair of specific chromosome. This is the normal condition, and abnormal occurrences are monosomy (q.v.), trisomy (q.v.) and nullisomy (with respectively one chromosome of a pair, three or none). There are also abnormal disomic conditions, such as when both chromosomes of the pair were inherited from the same parent.

Dispense Portion out a nutrient medium into containers, such as test tubes, jars, Erlenmeyer flasks, Petri dishes, etc.

Disaccharide A carbohydrate consisting of two linked sugar units.

Dissecting microscope A microscope with a low magnifying power of about 50×, used to examine or excise small plant or animal parts.

Dissection (L. *dissectio*, a dissecting or being dissected) Separation of a tissue by cutting for analysis or observation.

Dissolve Pass chemicals into solution.

Distillation (L. *distillatio*, a distilling process) The process of heating a mixture to separate the more volatile from the less volatile parts, and then cooling and condensing the resulting vapour so as to produce a more nearly pure or refined substance.

Di-sulphide bond A chemical bond that stabilizes the three-dimensional structure of proteins, and hence the protein's normal function. They form between cysteine residues in the same or different peptide molecules. a.k.a. di-sulphide bridge.

Diurnal Term describing the occurrence of an event at least once every 24 hours. *cf* circadian.

Dizygotic twins Two-egg twins, i.e., a pair of individuals that shared the same uterus at the same time, but which arose from separate and independent fertilization of two ova.

DNase (deoxyribonuclease) An enzyme that catalyses the cleavage of DNA. DNase I is a digestive enzyme secreted by the pancreas, that degrades DNA into shorter nucleotide fragments. Many other endonucleases and exonucleases are involved in DNA repair and replication. *cf* nuclease.

DNA bank In AnGR: Storage of DNA, which may or may not be the complete genome, but should always be accompanied by inventory information. *(Note:* at the present time, animals cannot be re-established from DNA alone.) (Source: FAO, 1999)

DNA construct A DNA molecule inserted into a cloning vector, usually a plasmid.

DNA diagnosis The use of DNA polymorphisms to detect the presence of a specific allele (often associated with a disease or syndrome) or DNA sequence.

DNA helicase (gyrase) An enzyme that catalyses the unwinding of the complementary strands of a DNA double helix.

DNA hybridization The pairing of two DNA molecules, often from different sources, by hydrogen bonding between complementary nucleotides. This technique is frequently used to detect the presence of a specific nucleotide sequence in a DNA sample.

DNA ligase An enzyme that catalyses a reaction that links two DNA molecules via the formation of a phospho-diester bond between the 3′ hydroxyl and 5′ phosphate of adjacent nucleotides. It plays an important role in DNA repair and replication. DNA ligase is one of the essential tools of recombinant DNA technology, enabling (among other things) the incorporation of foreign DNA into vectors. The ligase enzyme encoded by phage T4 is commonly used in gene-cloning experiments. It requires ATP as a co-factor. T4 is used *in vitro* to join the vector and insert DNAs.

Dolly The [name of the] first mammal to be created by cloning a cell from an adult animal. In this particular case, the cell came from the mammary tissue of an adult ewe. The creation of Dolly showed that the process of differentiation into adult tissue is not, as previously thought, irreversible. The result was achieved by nuclear transfer (q.v.). Dolly's birth was announced in 1997. Since then, cattle and mice have also been cloned from adult cells.

Domain A segment of a protein that has a discrete function or conformation. At the protein level, a domain can be as small as a few amino acid residues or as large as half of the entire protein.

Domestic animal diversity (DAD) In AnGR: The spectrum of genetic differences within each breed, and across all breeds within each domestic animal species, together with the species differences; all of which are available for the sustainable intensification of food and agriculture production.

(Source: FAO, 1999)

Dominance The type of gene action exhibited by a dominant allele (q.v.).

Dosage compensation A phenomenon whereby inactivation of all but one of the X chromosomes in female mammals results in males and females producing the same quantity of peptide from X-linked genes.

Double crossing-over Two simultaneous reciprocal breakage and reunion events between the same two chromatids.

Double helix Describes the coiling of the antiparallel strands of the DNA molecule, resembling a spiral staircase in which the paired bases form the steps and the sugar-phosphate backbones form the rails.

Double fertilization A process, unique to flowering plants, in which two male nuclei, which have travelled down the pollen tube, separately fuse with different female nuclei in the embryo sac. The first male nucleus fuses with the egg cell to form the zygote; the second male nucleus fuses with the two polar nuclei to form a triploid nucleus that develops into the endosperm.

Double recessive An organism homozygous for a recessive allele at each of two loci.

Double-stranded complementary DNA (dscDNA) A double-strand DNA molecule created from a cDNA template.

Dry weight The moisture-free weight of tissue obtained by drying at high (oven-drying) or low (freeze-drying) temperatures for an interval sufficient to remove all water.

Elastin A fibrous protein that is the major constituent of the yellow elastic fibres of connective tissue.

Electro-blotting The electrophoretic transfer of macromolecules (DNA, RNA or protein) from a gel, in which they have been separated, to a support matrix, such as a nitrocellulose sheet. A transfer usually used in techniques such as Southern and northern blotting.

Electromagnetic radiation Electromagnetic waves, including ultraviolet (UV), X-rays, and gamma radiation (g rays). Electromagnetic radiation is used to produce mutant cells or organisms, or, in the case of UV, disinfestation and sterilization, in tissue culture.

Electromagnetic spectrum The range of wavelengths or frequencies over which electromagnetic radiation extends.

Endogamy The fusion of reproductive cells from closely related parents, i.e., inbreeding. *cf* exogamy.

Endogenote The part of the bacterial chromosome that is homologous to a genome fragment (exogenote) transferred from the donor to the recipient cell in the formation of a merozygote.

Endogenous (Gr. *endon*, within, + *genos*, race, kind) Developed or added from within the cell or organism.

Endomitosis Duplication of chromosomes without division of the nucleus, resulting in increased chromosome number within a cell. Chromosome strands separate, but the cell does not divide.

Epiphyte A plant that grows upon another plant, but is neither parasitic on it nor rooted in the ground.

Episome A genetic extrachromosomal element (e.g., the fertility factor (F) in *Escherichia coli*) which replicates within a cell independently of the chromosome and is able to integrate into the host chromosome. The step of integration may be governed by a variety of factors and so the term episome has lost favour and been superseded by the wider term *plasmid*. Plasmids and F factors are episomes.

Epistasis Interaction between genes at different loci, e.g., one gene suppresses the effect of another gene that is situated at a different locus. Suppressed genes are said to be hypostatic. Dominance is associated with members of allelic pairs, whereas epistasis is interaction among products of non-alleles.

Epitope A specific chemical domain on an antigen that stimulates the production of, and is recognized by, an antibody. Each epitope on a molecule such as protein elicits the synthesis of a different antibody. a.k.a. antigenic determinant.

Epizootic A disease affecting a large number of animals simultaneously.

Equational division Mitotic-type division that is usually the second division in the meiotic sequence; somatic mitosis and the non-reductional division of meiosis. A chromosome division in which the two chromatids of each duplicated chromosome separate longitudinally, prior to being incorporated into two daughter nuclei.

Equatorial plate The figure formed by the chromosomes in the centre (equatorial plane) of the spindle in mitosis.

Equilibrium A state of dynamic systems in which there is no net change.

Equilibrium density gradient centrifugation A procedure used to separate macromolecules based on their density (mass per unit volume).

Factorial mating A mating scheme in which each male parent is mated with each female parent. Made possible in animals by means of *in vitro* embryo production (q.v.). Such a mating scheme substantially reduces the rate of inbreeding in a selection programme.

FAD (flavin adenine dinucleotide) A co-enzyme important in various biochemical reactions. It comprises a phosphorylated vitamin B_2 (riboflavin) molecule linked to the nucleotide adenine monophosphate (AMP). It functions as a hydrogen acceptor in dehydrogenation reactions, being reduced to $FADH_2$. This in turn is oxidized to FAD by the electron transport chain, thereby generating ATP (two molecules of ATP per molecule of $FADH_2$).

False-negative A negative assay result that should have been positive.

False-positive A positive assay result that should have been negative.

Farm animal genetic resources (AnGR) In AnGR: Those animal species that are used, or may be used, for the production of food and agriculture, and the populations within each of them. These populations within each species can be classified as wild and feral populations, landraces and primary populations, standardized breeds, selected lines, and any conserved genetic material. (Source: FAO, 1999)

Fed-batch fermentation Culture of cells or micro-organisms where nutrients are added periodically to the bioreactor.

Feedback inhibition The process by which the accumulated end product of a biochemical pathway stops synthesis of that product. A late metabolite of a synthetic pathway regulates synthesis at an earlier step of the pathway. *cf* end-product inhibition.

Fermentation The anaerobic breakdown by micro-organisms of complex organic substances, especially carbohydrates like glucose. The process is energy-yielding.

Fermentation is often misused to describe large-scale aerobic cell culture in specialized vessels (fermenters, bioreactors) for secondary product synthesis.

Fermentation substrates Materials used as food for growing micro-organism. The fermentation substrates and the trace materials needed, together with chemicals added to make the fermentation easier, form the culture medium.

Fitness The number of offspring left by an individual, often compared with the average of the population or with some other standard, such as the number of offspring left by a particular genotype.

Fixation An event that occurs when all the alleles at a locus except one are eliminated from a population. The remaining allele, now occurring with a frequency of 100%, is said to have been fixed.

Flagellum (pl: flagella; adj: flagellate) A whiplike organelle of locomotion in certain cells; locomotor structures in flagellate protozoa.

Flaming A technique for sterilizing instruments. The instrument is dipped in alcohol (usually 95% (v/v) ethanol) and then the alcohol on the instrument is ignited, thus heat-sterilizing the tool surface.

Flanking region The DNA sequences extending either side of a specific sequence.

Floccule Micro-organism aggregate or colloidal particle floating in or on a liquid. Contaminated liquid media are usually cloudy, illustrating this flocculation phenomenon.

Flocculant Agent that causes small particles to aggregate (flocculate).

Fluorescence *in situ* hybridization (FISH) Hybridization of cloned DNA to intact chromosomes, where the cloned DNA has been labelled with a fluorescent dye. This is the major method of physical mapping of cloned DNA fragments on chromosomes.

Fluorescence-activated cell sorting (FACS) The use of laser beams to detect differences in fluorescence between different types of cells in a mixture, and the subsequent deflection of cells into separate bins corresponding to each type of cell in the mixture. One of the popular uses of this technology is in sperm sexing (q.v.).

Fluorescent probe Probe whose response is based on the fluorescence intensity of individual cells or cell components.

Flow cytometry A technique used to sort cells or other biological materials by means of flow through apertures of defined size or by laser sorting.

Flower The structure in angiosperms (flowering plants) that bears the organs for sexual reproduction.

Foetus Pre-natal stage of a viviparous animal, between the embryonic stage and parturition.

Food processing enzyme Enzyme used to control food texture, flavour, appearance and, to a certain extent, nutritional value. Amylases break down complex polysaccharides to simplex sugars; proteases tenderize meat proteins. Biotechnology can assist the development of new food enzymes by finding or engineering enzymes which fit better with the other processes which the food must undergo, like cooking or canning.

Fog Fine particles of liquid suspended in the air, such as of water in a fog chamber used for acclimatizing recent *ex vitro* transplants. *cf* mist propagation.

Fold-back The structure formed when a double-stranded DNA molecule containing an inverted repeat sequence is denatured and then allowed to re-anneal at low DNA concentrations. The repeated sequence permits the formation of a double-stranded region within each of the separated strands of the original molecule.

Folded genome The condensed intracellular state of the DNA in the nucleoid of a bacterium. The DNA is segregated into domains, and each domain is independently negatively supercoiled.

Follicle Any enclosing cluster of cells that protects and nourishes a cell or structure within. Thus a follicle in the ovary contains a developing egg cell, while a hair follicle envelops the root of hair.

Follicle stimulating hormone (FSH) A hormone, secreted by the anterior pituitary gland in mammals, that stimulates, in female mammals, ripening of specialized structures in the ovary (Graafian follicles) that produce ova and, in males, the formation of sperm in the testis. It is a major constituent of fertility drugs.

Gram-positive organism Any prokaryotic organism that retains the first stain used in the Gram technique, which gives a purple-black colour when viewed under a light microscope. Retention of the stain is due to the structure of the cell wall.

Gram staining A technique to distinguish between two major bacterial groups, based on stain retention by their cell walls. Bacteria are heat-fixed, then stained with crystal violet, followed by iodine solution, and then rinsed with alcohol or acetone. Gram-positive bacteria are stained bright purple, while Gram-negative bacteria are decolourized.

Grana (L. *granum*, a seed) Structures within chloroplasts, seen as green granules with the light microscope and as a series of parallel lamellae with the electron microscope. Disc- or sac-like structures found in chloroplasts composed of stacked membranes and containing the chlorophyll and carotenoid pigments directly involved in photosynthesis (singular: granum).

Growth cabinet A cupboard used for incubating tubes or culture vessels under controlled environmental conditions. The degree of control over temperature, light and humidity is a function of the quality of the cabinet. *cf* culture room; incubator.

Growth factor Any of various chemicals, particularly polypeptides, that have a variety of important roles in the stimulation of new cell growth and cell maintenance. They bind to the cell surface on receptors. Specific growth factors can cause new cell proliferation.

Growth hormone (GH); **somatotrophin**; **somatotropin** A hormone, secreted by the mammalian pituitary gland, that stimulates protein synthesis and growth of the long bones in the legs and arms. It also promotes the breakdown and use of fats as an energy source, rather than glucose. Production of growth hormone is greatest during early life. Its secretion is controlled by the opposing actions of two hormones from the hypothalamus: somatocrinin (growth-hormone-releasing hormone), which promotes its release; and somatostatin (growth-hormone-inhibiting hormone), that inhibits the release of growth hormone from the anterior pituitary gland.

Growth inhibitor Any substance inhibiting the growth of an organism. The inhibitory effect can range from mild inhibition (growth retardation) to severe inhibition or death (toxic reaction). Two plant growth regulators that may act as inhibitors are ethylene and abscisic acid. The concentration of the inhibitor, the length of exposure to it, and the relative susceptibility of the organisms exposed to the inhibitor, are all important factors which determine the extent of the inhibitory effect.

Growth phases; growth phase curve The characteristic periods in the growth of a bacterial culture, as indicated by the shape of a graph of viable cell number versus time, namely: lag phase; logarithmic (or exponential) phase ; stationary phase; death phase.

Growth regulator A synthetic or natural compound that at low concentrations elicits and controls growth responses in a manner similar to hormones.

Growth retardant A chemical that selectively interferes with normal hormonal promotion of growth and other physiological processes, but without appreciable toxic effects.

Growth ring Any of the rings that can be seen in a cross-section of a woody stem, such as a tree trunk. It represents the xylem formed in one year as a result of fluctuating activity of the vascular cambium.

Growth substance Any organic substance, other than a nutrient, that is synthesized by plants and regulates growth and development. They are usually made in a particular region, such as the shoot tip, and transported to other regions, where they take effect.

Guanine A purine derivative that is one of the major component bases of nucleotides and the nucleic acids, DNA and RNA.

Guanosine A nucleoside consisting of one guanine molecule linked to a D-ribose sugar molecule. The derived nucleotides, guanosine mono-, di-, and triphosphate (GMP, GDP, and GTP, respectively), are important in various metabolic reactions.

Guide RNA An RNA molecule that contain sequences that function as a template during RNA editing.

Gymnosperm Any plant whose ovules and the seeds into which they develop are borne unprotected, rather than enclosed in ovaries, as are those of the flowering plants (the term gymnosperm means naked seed).

Gynandromorph An individual in which one part of the body is female and another part is male; a sex mosaic.

Gynogenesis Female parthenogenesis; after fertilization of the ovum, the male nucleus is eliminated and the haploid individual (described as gynogenetic) so produced possesses the maternal genome only. *cf* androgenesis; parthenogenesis; anther culture.

Habituation The phenomena that, after a number of sub-cultures, cells can grow, without the addition of specific factors, such as no longer needing exogenous growth regulators in the tissue culture medium. Such cells are autonomous.

Haemoglobin Conjugated protein compound containing iron, located in erythrocytes of vertebrates; important in the transportation of oxygen to the cells of the body.

Haemolymph The mixture of blood and other fluids in the body cavity of an invertebrate.

Hair A single or multicellular, absorptive (root hair) or secretory (glandular hair) and sometimes only a superficial outgrowth (covering hair) of the epidermal cells. The term trichome is often used but includes outgrowths from tissue below the epidermal layer. Distinguishing between hairs and trichomes can be difficult. Trichomes are usually connected to the vascular system, whereas hairs lack a vascular connection.

Hairy root culture A fairly recent development in plant culture, consisting of highly branched roots of a plant. A plant tissue is treated with a culture of the bacterium *Agrobacterium rhizogenes*, which transfers part of its own plasmid DNA to the cells of an infected plant. This alters the plant's metabolism, including alterations in hormone levels, which in turn cause the explant to grow highly branched roots from the sites of infection. The roots branch much more frequently than the usual root system of that plant, and are also covered with a mass of tiny root hairs. Their most significant feature is that they produce secondary metabolites at levels similar to those made in the original plant. Thus they can be used as replacement plants for making such compounds as food flavours or fragrances.

Haploid cell A cell containing only one set, or half the usual (diploid) number, of chromosomes.

Haplotype 1. A group of alleles, each from a different locus in the same region of a chromosome, that exist in the same double helix.

Haptoglobin A serum protein, alpha globulin, that interacts with haemoglobin during recycling of the iron molecule of haemoglobin.

Hardening off Adapting plants to outdoor conditions by gradually withholding water, lowering the temperature, increasing light intensity, or reducing the nutrient supply. The hardening-off process conditions plants for survival when transplanted outdoors. The term is also used for gradual acclimatization to *in vivo* conditions of plants grown *in vitro*, e.g., gradual decrease in humidity. *cf* acclimatization; free-living conditions.

Herbicide Any substance that is toxic to plants; usually applied to agrochemicals intended to kill specific unwanted plants, such as weeds.

Herbicide resistance The ability of a plant to withstand herbicide. Herbicide resistance has been one of the early targets of plant genetic engineering. If a herbicide is sprayed onto a field planted with such resistant crops, then all the plants except the crop would be killed, thus providing an effective method of weed control without having to develop herbicides specific to each weed type. There is substantial concern in some quarters about the widespread use of this technology, which is essentially giving the plant kingdom the ability to evade man's most effective herbicides. The concern are that, firstly, such engineering will lead to increased use of the herbicides, at a time when it is generally accepted that the use of chemicals should be kept as low as possible, and that, secondly, there is the possibility that resistant crop plants will escape to become weeds, or that their resistant genes could be transferred to other species, including weeds.

Heredity Resemblance among individuals related by descent; transmission of traits from parents to offspring.

Heteroduplex A double-stranded DNA molecule or DNA-RNA hybrid, where each strand is of a different origin, and consequently containing one or more mismatched (non-complementary) base pairs. A DNA duplex is prepared by the hybridization of single-stranded DNA molecules derived from two different sources. Where the two DNAs have identical or very similar sequences, a double-stranded molecule will be established, whereas where the two DNAs differ in sequence, single-stranded regions will remain. A heteroduplex will be revealed as single-strand "bushes" when DNA is observed electron microscopically. A map of homologous and non-homologous regions of the two molecules may thereby be constructed. This process is known as heteroduplex mapping.

Heterogametic sex Producing unlike gametes with regard to the sex chromosomes. In mammals, the XY male is heterogametic, and the XX female is homogametic.

Heterokaryon Cell with two or more different nuclei as a result of cell fusion. Opposite: homokaryon.

Heterologous From a different source, as in heterologous DNA.

Heterologous probe A DNA probe that is derived from one species and used to screen for a similar DNA sequence from another species.

Heteroplasmy A cellular condition in which two genetically different types of organelles are present. *cf* homoplasmy.

Heteroploid Term given to a cell culture when the cells comprising the culture possess nuclei containing chromosome numbers other than the diploid number.

Heteropyknosis (adj: heteropyknotic) Property of certain chromosomes, or of their parts, to remain more dense during the cell cycle and to stain more intensely than other chromosomes or parts.

Heterozygote (adj: heterozygous) (Gr. *heteros*, different + *zygon*, yoke) An individual that has different alleles at the same locus in its two homologous chromosomes.

Hfr High-frequency recombination strain of *Escherichia coli;* in such strains, the F episome is integrated into the bacterial chromosome.

Histocompatibility The degree to which tissue from one organism will be tolerated by the immune system of another organism.

Histocompatibility complex; histocompatibility system The collection of genes coding for peptides present on the surface of nucleated cells; these peptides are responsible for the differences between genetically non-identical individuals that cause rejection of tissue grafts between such individuals. These peptides were originally called histocompatibility antigens. They are now called histoglobulins, reflecting their structural similarity to immunoglobulins.

Histology (Gr. *histos*, cloth, tissue + *logos*, discourse) Science that deals with the microscopic structure of animal and plant tissues.

Histone Group of water-soluble proteins rich in basic amino acids, closely associated with DNA in plant and animal chromatin. Histones are involved in the coiling of DNA in chromosomes and in the regulation of gene activity.

Homodimer A protein with two identical polypeptide chains.

Homogenotization A genetic technique used to replace one copy of a gene, or other DNA sequence within a genome, with an altered copy of that sequence. The DNA is first cloned and then altered in some way, e.g., a transposon is inserted into a gene. The mutated gene copy can be used to replace the original gene by recombination *in vivo*. The incorporation of the mutated gene is usually selected, for example, by virtue of its containing a transposon-encoded antibiotic resistance.

Homokaryon Cell with two or more identical nuclei as a result of fusion. Opposite: heterokaryon.

Homologous From the same source, or having the same evolutionary function or structure.

Homologous chromosomes Chromosomes that occur in pairs and are generally similar in size and shape: one comes from the male parent and the other from the female. Such chromosomes contain the same linear array of genes.

Homologous recombination The exchange of DNA fragments between two DNA molecules or chromatids of paired chromosomes (during crossing over) at the site of identical nucleotide sequences.

Homology The degree of identity between individuals, or characters. The degree of identity between the nucleotide sequences of two nucleic acid molecules or the amino acid sequences of two protein molecules. Although sequence determination is the ultimate test of homology, useful estimates can be provided by either DNA-DNA or DNA-RNA hybridization.

Hormone (Gr. *hormaein*, to excite) A specific organic product, produced in one part of a plant or animal body, and transported to another part where, at low concentrations, it promotes, inhibits or quantitatively modifies a biological process.

Host-specific toxin A metabolite produced by a pathogen which has a host specificity equivalent to that of the pathogen. Such toxins are utilized for *in vitro* selection experiments to screen for tolerance or resistance to the pathogen.

Host An organism that contains another organism, or a cloning vector.

Hydrophobic interactions Association of non-polar groups with each other when present in aqueous solutions because of their insolubility in water.

Hydroponics The growing of plants in aerated water containing all the essential mineral nutrients, with no soil. Also called soilless gardening or cultivation.

3′-hydroxyl end The hydroxyl group that is attached to the 3′ carbon atom of the sugar (ribose or deoxyribose) of the terminal nucleotide of a nucleic acid molecule.

Hyperploid A genetic condition in which a chromosome or a segment of a chromosome is over-represented in the genotype. Opposite: hypoploid.

Hypersensitive sites Regions in the DNA that are highly susceptible to digestion with endonucleases.

Hypertonic A solution with an osmotic potential greater than that of living cells, leading to water loss from, shrinkage or plasmolysis of cells in a hypertonic situation. Opposite: hypotonic.

Hypervariable region The parts of both the heavy and light chains of an antibody molecule that enable it to bind to a specific site on an antigen.

Hypervariable segment A region of a protein that varies considerably between strains or individuals.

Hypochlorite Generic term for aqueous solutions of sodium hypochlorite, potassium hypochlorite or calcium hypochlorite, which are oxidizing agents and used for disinfecting surfaces and surface-sterilizing tissues, and for bleaching.

Hypocotyl (Gr. *hypo*, under + *kotyledon*, a cup-shaped hollow) Portion of an embryo or seedling below the cotyledons, which is a transitional area between stem and root.

Hypomorph A mutation that reduces but does not completely abolish gene expression.

Hypoplastic Reduction in plant growth or development (dwarfing, stunting) resulting from an abnormal condition associated with a disease or nutritional stress.

Hypoploid A genetic condition in which a chromosome or segment of a chromosome is underrepresented in the genotype. Opposite: hyperploid.

Hypothalamic peptides Peptides generated in the vertebrate forebrain and concerned with regulating the body's physiological state.

Hypothesis (Gr. *hypothesis*, foundation) A tentative theory or supposition provisionally adopted to explain certain facts and to guide in the investigation of other facts. Once proven by rigorous scientific investigation, it becomes a theory or a law.

Hypotonic Osmotic potential less than that of living cells. Cells placed in a hypotonic solution display swelling and turgidity.

Illuminate To supply or brighten with light. Illumination is an absolute requirement for tissue cultures. Fluorescent lights are commonly employed. The light intensity is dependent on the light source and the requirements of the culture.

Imaginal disc A mass of cells in the larvae of *Drosophila* and other holometabolous insects that gives rise to particular adult organs, such as antennae, eyes or wings.

Immediate early gene A viral gene that is expressed promptly after infection.

Immobilized cells Cells entrapped in matrices such as alginate, polyacrylamide and agarose designed for use in membrane and filter bioreactors.

Immortalization The genetic transformation of a cell type into a cell line which can proliferate indefinitely.

Immortalizing oncogene A gene that upon transfection enables a primary cell to grow indefinitely in culture.

Immune response The processes, including the synthesis of antibodies, that are used by vertebrates to respond to the presence of a foreign antigen.

Immunity The state of relative insusceptibility of an animal or plant to infection by disease-producing organisms or to the harmful effects of their poisons.

Immunization The production of immunity in an individual by artificial means. Active immunization involves the introduction, either orally or by infection, of specially treated bacteria, viruses or their toxins so as to stimulate the production of antibodies.

Immuno-affinity chromatography A purification technique in which an antibody is bound to a matrix and is subsequently used to bind and separate a protein from a complex mixture. *cf* affinity chromatography.

Immunoglobulin One of a group of proteins (globulins) in the body that act as antibodies. They are produced by specialized cells (B lymphocytes) and are present in blood serum and other body fluids.

Immunosuppression The suppression of immune response. Immunosuppression is necessary following organ transplants in order to prevent the host rejecting the grafted organ.

Immunoassay; immunodiagnostics An assay system which detects proteins by using an antibody specific to that protein. A positive result is seen as a precipitate of an antibody-protein complex. The antibody can be linked to a radioactive atom or to an enzyme which catalyses an easily monitored reaction such as a change in colour.

Immunochemical control Use of immune agents to combat infections.

Immunogenicity The ability to elicit an immune response.

Immunosensor A biosensor having an antibody as biological part.

Immunosuppressor A substance, an agent or a condition that prevents or diminishes the immune response.

Immunotherapy The use of an antibody or a fusion protein containing the antigen binding site of an antibody to cure a disease or enhance the well-being of a patient.

Immunotoxin Protein drugs consisting of an antibody joined to a toxin molecule. Immunotoxins can be made by linking toxin and antibody molecules chemically, or by fusing the genes for the toxin and the antibody. The antibody portion of the molecule facilitates binding to a target molecule or cell, and the toxin inactivates or kills the target molecule or cell.

Impeller An agitator that is used for mixing the contents of a bioreactor.

Inactivated agent A virus, bacterium or other organism that has been treated to prevent it from causing a disease.

Inbred line The product of inbreeding, i.e., the mating of individuals that have ancestors in common; in plants and laboratory animals, it refers to populations resulting from at least 6 generations of selfing or 20 generations of brother-sister mating, that are for all practical purposes, completely homozygous. In farm animals, the term is sometimes used to describe populations that have resulted from several generations of the mating of close relatives, without having reached complete homozygosity.

Inbreeding Matings between individuals that have one or more ancestors in common, the extreme condition being self-fertilization, which occurs in many plants and some primitive animals.

Inbreeding depression Reduction in vigour, yield, etc., of a population that is commonly seen as the level of inbreeding increases. The traits that show greatest inbreeding depression are those that are most closely associated with viability and reproductive ability.

Inclusion body Protein that is overproduced in a recombinant bacterium and forms a crystalline array inside the bacterial cell.

Incompatibility group A classification scheme indicating which plasmids can co-exist within a single cell. Plasmids must belong to different incompatibility groups to co-exist within the same cell. Plasmids that belong to the same incompatibility group are unstable when placed in the same cell. A plasmid cloning vector should always belong to an incompatibility group different from that of the host bacterium's endogenous plasmids.

Incomplete dominance A type of gene action in which heterozygotes have a phenotype that is distinctly different from, and intermediate to, the homozygous phenotypes.

Incomplete penetrance When some individuals in a population have a specific genotype that causes an abnormality but are not affected.

Incubator An apparatus in which environmental conditions (light, photoperiod, temperature, humidity, etc.) are fully controlled, and used for hatching eggs, multiplying micro-organisms, culturing plants, etc. *cf* culture room; growth cabinet.

Indehiscent Describing a fruit or fruiting body that does not open to release its seeds or spores when ripe.

Independent assortment The random distribution of alleles (from different loci) to the gametes that occurs when genes are located in different chromosomes or far apart on large chromosomes. The distribution of alleles at one locus is independent of other the distribution of alleles at another locus. *cf* linkage.

Indirect embryogenesis Embryo formation on callus tissues derived from zygotic or somatic embryos, seedling plant or other tissues in culture. *cf* direct embryogenesis.

Indirect organogenesis Organ formation on callus tissues derived from explants.

Inducer A low-molecular-weight compound or a physical agent that is bound by a repressor so as to produce a complex that can no longer bind to the operator; thus, the presence of the inducer turns on the expression of the gene(s) controlled by the operator.

Inducible A gene or gene-product is said to be inducible if its transcription or synthesis is increased by exposure of the cells to an effector. Effectors are usually small molecules whose effects are specific to particular operons or groups of genes. *See* constitutive.

Inducible enzyme An enzyme that is synthesized only in the presence of the substrate that acts as an inducer.

Inducible gene A gene that is expressed only in the presence of a specific metabolite, the inducer.

Kanamycin An antibiotic of the aminoglycoside family that poisons translation by binding to the ribosomes.

Karyogram A diagram representing the characteristic features of the chromosomes of a species.

K_{cat} The catalytic rate constant that characterizes an enzyme-catalysed reaction. The larger the k_{cat} value, the faster the conversion of substrate into product.

K_{cat}/K_m The catalytic efficiency of an enzyme-catalysed reaction. The greater the value of k_{cat}/K_m, the more rapidly and efficiently the substrate is converted into product.

Killer T cells T cells that carry T-cell receptors and kill cells displaying the recognized antigens.

Kilobase (kb) A length unit equal to 1 000 base pairs of a double-stranded nucleic acid molecule. One kilobase of double-stranded DNA has a mass of about 660 kilodalton.

Kilobase pairs (kb) Used interchangeably with kilobase.

Kinase An enzyme that can transfer a phosphate from a high energy phosphate such as ATP, to an organic molecule.

Kinetics Dynamic processes involving motion.

Kinetin (Gr. *kinetikos*, causing motion) One of the cytokinins, a group of growth regulators that characteristically promote cell division in plants.

Kinetochores Two parallel bushlike filaments situated in the centromere of a chromosome.

Kinetosome Granular cytoplasmic structure which forms the base of a cilium or flagellum. *cf* basal body.

Kinin A substance promoting cell division. In plant systems, the prefix cyto- has been added (cytokinin) to distinguish it from kinin in animal systems.

Klenow fragment A product of proteolytic digestion of the DNA polymerase I from *E. coli*; it has both polymerase and 3′-exonuclease activities but not 5′-exonuclease activity.

K_m A dissociation constant that characterizes the binding of an enzyme to a substrate. The smaller the value of K_m, the tighter the binding of the enzyme to the substrate. Also called the Michaelis constant.

Knockout An animal resulting from an embryonic stem cell in which a normal functional gene has been replaced by a non-functional form of the gene. This technique is used extensively in mice - much can be learned about the function of a gene by studying the phenotype of animals that lack the peptide product of that gene.

Label A compound or atom that is either attached to or incorporated into a macromolecule and is used to detect the presence of a compound, substance, or macromolecule in a sample. a.k.a. tag.

Leptonema (adj: leptotene) Stage in meiosis immediately preceding synapsis, in which the chromosomes appear as single, fine, threadlike structures (but they are really double because DNA replication has already taken place).

Live vaccine A living, non-virulent form of a micro-organism or virus that is used to elicit an antibody response that will protect the inoculated organism against infection by a virulent form of the micro-organism or virus. Also a living, non-virulent micro-organism or virus that express a foreign antigenic protein and is used to inoculate humans or animals. The latter organisms are also called a live recombinant vaccines.

Locus (pl: loci) A site on a chromosome.

Lod score The logarithm of the odds of linkage between two loci. Calculated from pedigree data, as the log (to base 10) of the ratio of the probability of the observed pedigree assuming linkage with a specified recombination fraction q, to the probability of the observed pedigree assuming no linkage, i.e., recombination fraction = 0.5. A lod score (also called a *z* value) is thus calculated as $z = \log_{10}$ {Prob(data | q)/Prob(data | 0.5)}. A lod score of +3.00 (which is odds of 1000:1) or greater is regarded as acceptable evidence for linkage; -2.00 (which is the log of 1:100) or less indicates that no linkage exists.

Lymphokine Generic name for proteins that are released by lymphocytes to act on other cells involved in the immune response. The term includes interleukins and interferons.

Lymphoma Cancer originating in the lymph nodes, spleen and other lympho-reticular sites.

Lyase Any of a class of enzymes that catalyse either the cleavage of a double bond and the addition of new groups to a substrate, or the formation of a double bond.

Lyophilize Rapid freezing followed by a dehydration under high vacuum. The process is lyophilization. *cf* freeze-drying.

Lysis (Gr. *lysis*, a losing) The destruction or breakage of cells either by viruses or by chemical or physical treatment.

Lysogen A bacterial cell whose chromosome contains integrated viral DNA.

Lysogenic bacteria Bacteria harbouring temperate (non-virulent, lysogenic) bacteriophages.

Lysogeny A condition in which a bacteriophage genome (pro-phage) survives within a host bacterium, either as part of the host chromosome or as part of an extrachromosomal element, and does not initiate lytic functions.

Lytic A phase of the virus life cycle during which the virus replicates within the host cell, releasing a new generation of viruses when the infected cell lyses.

Lytic cycle The steps in viral production that usually lead to cell lysis.

Lysosome A membrane-bound sac within the cytoplasm of animal cells that contains enzymes responsible for the digestion of material in food vacuoles, the dissolution of foreign particles entering the cell and, on the death of the cell, the breaking down of all cell structures. The digestive system of the cell.

M13 A single-stranded DNA bacteriophage used as a vector for DNA sequencing.

M13 strand The single-stranded DNA molecule that is present in the infective form of bacteriophage M13.

Macerate To disintegrate tissues to obtain a cell dissociation. Cutting, soaking or enzymatic actions are commonly used.

Macromolecule Molecule of large molecular weight, such as proteins, nucleic acids and polysaccharides.

Macronutrient (Gr. *makros*, large + L. *nutrire*, to nourish) For growth media: an essential element normally required in concentrations >0.5 millimole/l.

Macrophages Large, white blood cells that ingest foreign substances and display on their surfaces antigens produced from the foreign substances, to be recognized by other cells of the immune system.

Macropropagation Production of plant clones from growing parts.

Major histocompatibility complex The large cluster of genes that encode the major histocompatibility antigens in mammals.

Malignant Having the properties of cancerous growth.

Management of farm animal genetic resources In AnGR: The sum total of technical, policy and logistical operations involved in understanding (characterization), using and developing (utilization), maintaining (conservation), accessing, and sharing the benefits of animal genetic resources. (Source: FAO, 1999)

Mannitol ($C_6H_{14}O_6$; f.w. 182.17) A sugar alcohol widely distributed in plants. Mannitol is commonly used as a nutrient and osmoticum (q.v.) in suspension medium for plant protoplasts.

Microbody (Gr. *mikros*, small + body) A cellular organelle always bound by a single membrane, frequently spherical, from 20 to 60 nm in diameter, containing a variety of enzymes.

Micro-carriers Small particles used as a support material for cells, and particularly mammalian cells, which are too fragile to be pumped and stirred as bacterial cells are in a large-scale culture.

Microdroplet array; multiple drop array (MDA); hanging droplet technique. Introduced by Kao and Konstabel (1970), this technique is used to evaluate large numbers of media modifications, employing small quantities of medium into which are placed small numbers of cells. Droplets of liquid are arranged on the lid of a Petri dish, inverted over the bottom half of the dish containing a solution with a lower osmotic pressure, and the dish is sealed. The cells or protoplasts form a monolayer at the droplet meniscus and can easily be examined.

Micro-element An element required in very small quantities.

Micro-encapsulation A process of enclosing a substance in very small sealed capsules from which material is released by heat, solution or other means.

Micro-environment (Gr. *mikros*, small + O.F. *environ*, about) The environment close enough to the surface of a living or non-living object to be influenced by it.

Microfibrils (Gr. *mikros*, small + *fibrils*, diminutive of fibre) Microfibrils are exceedingly small fibres visible only at the high magnification of the electron microscope.

Micro-injection The introduction of small amounts of material (DNA, RNA, enzymes, cytotoxic agents) into a single eukaryotic cell with a fine, microscopic needle, penetrating the cell membrane.

Micro-isolating system Mechanical separation of single cells or protoplasts thus allowing them to proliferate individually.

Necrosis Death associated with discoloration and dehydration of all or some parts of organs.

Negative autogenous regulation; negative self-regulation Inhibition of the expression of a gene or set of coordinately regulated genes by the product of the gene or the product of one of the genes.

Negative control system A mechanism in which the regulatory protein(s) is required to turn off gene expression.

Nematodes A class of slender, unsegmented worms, often parasitic. a.k.a. eelworms, especially when phytoparasitic.

Neo-formation Organogenesis; production of newly formed structures, such as tissues, meristems and embryos.

Neoplasm Localized cell multiplication. Generally it designates a collection of cells which have undergone genetic transformation, forming a tumour. Neoplasmic cells differ in structure and function from the original cell type.

Neoteny The retention of juvenile body characters in the adult state, or the occurrence of adult characters in the juvenile state.

Net photosynthesis Photosynthetic activity minus respiratory activity, as measured by carbon dioxide exchange.

Neutral mutation A mutation that changes the nucleotide sequence of a gene but has negligible effect on the fitness of the organism.

Neutral theory The theory that much of evolution has been primarily due to random drift of neutral mutations.

Nick Verb: To break a phospho-diester bond in the backbone of one of the strands of a duplex DNA molecule. *cf* cleave; cut.

Nicked circle; relaxed circle During extraction of plasmid DNA from the bacterial cell, one strand of the DNA becomes nicked. This relaxes the torsional strain needed to maintain supercoiling, producing the familiar form of plasmid.

Nick translation A procedure for labelling DNA. A DNA fragment is treated with DNase to produce single-stranded nicks. The nick is moved along the DNA molecule in the presence of labelled deoxyribonucleoside triphosphates by the concerted action of the 5′-> 3′ exonuclease and 5′-> 3′ polymerase activities of *E. coli* DNA polymerase I.

Nitrification A chemical process in which nitrogen in plant and animal wastes and dead remains is oxidized, first to nitrites and then to nitrates.

Nitrate The only form in which nitrogen can be used directly by plants; a component of chemical fertilizers.

Nitrite A salt or ester of nitrous acid.

Nitrocellulose; cellulose nitrate A nitrated derivative of cellulose. It is made into membrane filters of defined porosity, used to immobilize DNA, RNA or protein, which can then be probed with a labelled sequence or antibody. These filters have a variety of uses in molecular biology, particularly in nucleic acid hybridization experiments. Used extensively in the Southern and northern blotting procedures involving DNA and RNA.

Nuclein The term used by Friedrich Miescher to describe the nuclear material he discovered in 1869, which today is known as DNA.

Nucleo-cytoplasmic ratio In a cell, the ratio of nuclear to cytoplasmic volume. This ratio is high in meristematic cells and low in differentiated cells.

Nucleolar organizer (NO); **nucleolar organizer region** (NOR) A chromosomal segment containing genes that encode ribosomal RNA; located at the secondary constriction of some chromosomes.

Nucleolar zone Any chromosome region, irrespective of whether or not it is a secondary-constriction, that is associated with the formation of the nucleolus during telophase.

Nucleolus (L. *nucleolus*, a small nucleus) An RNA-rich intranuclear organelle in the nucleus of eukaryotic cells, produced by a nucleolar organizer. It represents the storage place for ribosomes and ribosome precursors. The nucleolus consists primarily of ribosomal precursor RNA, ribosomal RNA, their associated proteins, and some, perhaps all, of the enzymatic equipment (RNA polymerase, RNA methylase, RNA cleavage enzymes) required for synthesis, conversion and assembly of ribosomes. Subsequently the ribosomes are transported to the cytoplasm.

Nucleoplasm The non-staining or slightly chromophilic, liquid or semi-liquid, ground substance of the interphase nucleus and which fills the nuclear space around the chromosomes and the nucleoli. Little is known of the chemical composition of this ground substance, which is not easily defined. It may be called "karyoplasm" when it is gel-like, and "karyolymph" when it is a colloidal fluid, but generally the terms are synonymous.

Nucleoprotein Conjugated protein composed of nucleic acid and protein; the material of which the chromosomes are made.

Nucleoside A base (purine or pyrimidine) that is covalently linked to a 5-carbon (pentose) sugar. When the sugar is ribose, the nucleoside is a ribonucleoside; when it is deoxyribose, the nucleoside is a deoxyribonucleoside. Adenine, guanine and cytosine occur in both DNA and RNA; thymine occurs in DNA; and uracil in RNA. They are the building blocks of DNA and RNA.

Nucleoside analogue A synthetic molecule that resembles a naturally occurring nucleoside, but that lacks the bond site needed to link it to an adjacent nucleotide.

Nucleosome Spherical sub-units of eukaryotic chromatin that are composed of a core particle consisting of an octamer of histones (two molecules each of histones H_{2a}, H_{2b}, H_3 and H_4) and 146 nucleotide pairs.

Nucleotide A nucleoside with one or more phosphate groups linked to the 5′ carbon of the pentose sugar. Ribose-containing nucleosides include ribonucleoside monophosphate (NMP), ribonucleoside diphosphate (NDP), and ribonucleoside triphosphate (NTP). When the nucleoside contains the sugar deoxyribose, the nucleotides are called deoxyribonucleoside mono-, di-, or tri-phosphates (dNMP, dNDP, or dNTP). A building block of DNA and RNA.

Nucleus (L. *nucleus*, kernel of a nut) A dense protoplasmic-membrane-bound region of a eukaryotic cell that contains the chromosomes separated from the cytoplasm by a membrane; present in all eukaryotic cells except mature sieve-tube elements.

Nullisomy (adj: nullisomic) An otherwise diploid cell or organism lacking both members of a chromosome pair (chromosome formula $2n$ -2).

Nurse culture Planting a cell from a suspension culture on a raft of filter paper above a callus tissue piece (nurse tissue). The filter paper serves to prevent tissue union but allows the flow of essential substances from the nurse to the isolated cell. In such culture, a piece of callus is first placed on nutrient agar. Over this tissue is laid a strip of filter paper.

Nutrient cycle The passage of a nutrient or element through an ecosystem, including its assimilation and release by various organisms and its transformation into various organic or inorganic chemical forms.

Nutrient deficiency Absence or insufficiency of some factor needed for normal growth and development.

Nutrient film technique (NFT) Hydroponic technique used to grow plants. NFT delivers a film of water or nutrient solution either continuously or through on-off cycles (e.g., on 8 minutes and off 7 minutes).

Nutrient gradient A diffusion gradient of nutrients and gases that develops in tissues where only a portion of the tissue is in contact with the medium. Gradients are less likely to form in liquid media than in callus cultures.

Nutrient medium (pl: nutrient media) A solid, semi-solid or liquid combination of: major and minor salts; an energy source (sucrose); vitamins; plant growth regulators; and occasionally other defined or undefined supplements. Often made from stock solutions, then sterilized by autoclaving or filtering through a micropore filter.

Octoploid Cell or organism with eight sets of chromosomes, i.e., chromosome number $2n = 8x$.

Oestrogen; estrogen The generic term for a group of female sex hormones which control the development of sexual characteristics and control oestrus.

Oestrous cycle (from oestrus) The cycle of reproductive activity shown by most sexually mature non-pregnant female mammals.

Oestrus (adj: oestrous) In female mammals, the period of sexual excitement and acceptance of the male. a.k.a. rut, heat.

Offset Young plant produced at the base of a mature plant.

Offshoot Short, usually horizontal, stem produced near the crown of a plant.

Offspring; progeny (both same in plural) New individual organisms that result from the process of sexual or asexual reproduction.

Okazaki fragment Since DNA can be replicated in only one direction, i.e., nucleotides can be added only at the 3′ end, only one of the two strands of a double helix can be replicated continuously. The other strand is replicated in small segments (Okazaki fragments) that are subsequently joined together by DNA ligase.

Oligomer A molecule formed from a small number of monomers.

Oligonucleotide A short molecule (usually 6 to 100 nucleotides) of single-stranded DNA.

Oligosaccharide Carbohydrate consisting of several linked sugar units.

Oncogene A gene that causes cells to grow in an uncontrolled manner, i.e., that causes cancer. Oncogenes are mutant forms of normal functional genes (called proto-oncogenes) that have a role in normal cell proliferation. Oncogenes are found in tumours and in retroviruses; in the latter case, having been picked up along with retroviral genes during retroviral replication in the host. Once incorporated as part of the retroviral genome, they are activated at inappropriate times and places by retroviral promoters, thereby becoming oncogenes.

Onco-mouse A mouse that has been genetically modified to incorporate an oncogene; oncogenes cause cells to undergo cancerous transformation.

Ontogeny Developmental life history of an organism.

Oocyte The egg mother cell; it undergoes two meiotic divisions (oogenesis) to form the egg cell. The primary oocyte is before completion of the first meiotic division; the secondary oocyte is after completion of the first meiotic division.

Oogenesis The formation and growth of the egg or ovum in an animal ovary.

Open pollination Pollination by wind, insects or other natural mechanisms.

Open reading frame (ORF) A sequence of nucleotides in a DNA molecule that has the potential to encode a peptide or protein: it starts with a start triplet (ATG), is followed by a string of triplets each of which encodes an amino acid, and ends with a stop triplet (TAA, TAG or TGA). This term is often used when, after the sequence of a DNA fragment has been determined, the function of the encoded protein is not known. The existence of open reading frames is usually inferred from the DNA (rather than the RNA) sequence.

Operational definition An operation or procedure that can be carried out to define or delimit something.

Operator The region of DNA that is upstream from a gene or genes and to which one or more regulatory proteins (repressor or activator) binds to control the expression of the gene(s).

Operon A functionally integrated genetic unit for the control of gene expression in bacteria. It consists of one or more genes that encode one or more polypeptide(s) and the adjacent site (promoter and operator) that controls their expression by regulating the transcription of the structural genes.

Opine The condensation product of an amino acid with either a keto-acid or a sugar, produced by the crown gall tissues. Opine synthesis is a unique characteristic of tumour cells.

Ordinate The vertical axis of a graph. Opposite: absciss; abscissa.

Organ A tissue or group of tissues that constitute a morphologically and functionally distinct part of an organism.

Organ culture The aseptic culture of complete living organs of animals and plants outside the body in a suitable culture medium. Animal organs must be small enough to allow the nutrients in the culture medium to penetrate all the cells.

Organellar genes Genes located on organelles outside the nucleus.

Organelle A membrane-bounded specialized region within a cell, such as the mitochondrion or dictyosome, that carries out a specialized function in the life of a cell.

Organic Referring in chemistry to compounds containing carbon, many of which have been in some manner associated with living organisms.

Organic complex A chemically undefined compound added to nutrient media to stimulate growth, e.g., coconut milk; yeast extract; casein hydrolysate.

Organic co-solvent A compound used to dissolve some neutral organic substances, such as in media preparation. Organic co-solvents include alcohols (usually ethanol), acetone and dimethylsulphoxide (DMSO).

Organic evolution The process by which changes in the genetic composition of populations of organisms occur in response to environmental changes.

Organized tissue Composed of regularly differentiated cells.

Organizer An inductor; a chemical substance in a living system that determines the fate in development of certain cells or groups of cells.

Organoid An organ-like structure produced in culture, such as leaves, roots or callus.

Organoleptic Having an effect on one of the organs of sense, such as taste or smell.

Organism An individual living system, such as animal, plant or micro-organism, that is capable of reproduction, growth and maintenance.

Origin of replication The nucleotide sequence at which DNA synthesis (replication) is initiated.

Ortet The plant from which a clone is obtained.

Osmic acid (= osmium tetroxide) (OsO_4; f.w. 254.20) A fixing agent commonly used to prepare tissue samples for electron microscopy.

Osmolarity The total molar concentration of the solutes. Osmolarity affects the osmotic potential of solution or nutrient medium.

Osmosis (Gr. *osmos*, a pushing) Diffusion from areas of high concentration to areas of lower concentration of a solvent through a differentially permeable membrane.

Osmotic potential [value] Potential brought about by dissolving a substance, especially in water.

Osmoticum An agent, such as PEG, mannitol, glucose or sucrose, employed to maintain the osmotic potential of a nutrient medium equivalent to that of the cultured cells (isotonic). Because of this osmotic equilibrium, cells are not damaged *in vitro*.

Outbreeding A mating system characterized by the breeding of genetically unrelated or dissimilar individuals. Since genetic diversity (q.v.) tends to be enhanced and since vigour or fitness of individuals can be increased by this process, it is often used to counter the detrimental effects of continuous inbreeding.

Outflow The volume of growing cells that is removed from a bioreactor during a continuous fermentation process.

Oxygen-electrode-based sensor Sensor in which an oxygen electrode - a standard electrochemical cell which measures the amount of oxygen in a solution - is coated with a biological material which generates or absorbs oxygen. When the biological coating is active, the amount of oxygen next to the electrode changes and the signal from the electrode changes.

Palaeontology The study of the fossil record of past geological periods and of the phylogenetic relationships between extinct and contemporary plant and animal species.

Palisade parenchyma Elongated cells found just beneath the upper epidermis of leaves, and containing many chloroplasts.

Panicle (L. *panicula*, a tuft) An inflorescence, the main axis of which is branched; the branches bear loose racemose flower clusters.

Panicle culture Aseptic culture of grain panicle segments to induce microspore germination and development.

Panmictic population A population in which mating occurs at random.

Panmixis Random mating in a population. *cf* apomixis.

Paracentric inversion An inversion that is entirely within one arm of a chromosome and does not include the centromere.

Paraffin [wax] A translucent, white, solid hydrocarbon with a low melting point. Paraffin is used as an embedding medium to support tissue for sectioning for light microscopy observation.

Parallel evolution The development of different organisms along similar evolutionary paths due to similar selection pressures acting on them.

ParafilmTM A stretchable film based on paraffin wax; used to seal tubes and Petri dishes. ParafilmTM is a proprietary name which is applied colloquially to similar products.

Parahormone A substance with hormone-like properties that is not a secretory product (e.g., ethylene; carbon dioxide).

Parameter A value or constant pertaining to an entire population. *cf* statistic.

Parasexual cycle A sexual cycle involving changes in chromosome number but differing in time and place from the usual sexual cycle; occurring in those fungi in which the normal cycle is suppressed or apparently absent.

Parasite (Gr. *parasites*, one who eats at the table of another) An organism deriving its food from the living body of another organism.

Parasitism The close association of two or more dissimilar organisms, where the association is harmful to at least one. *cf* commensalism; symbiosis.

Parasporal crystal Tightly packaged insect pro-toxin molecules that are produced by strains of *Bacillus thuringiensis* during the formation of resting spores.

Penetrance The percentage of individuals in population that show a particular phenotype among those capable of showing it, i.e., among those that have the genotype normally associated with that phenotype.

Peptide A sequence of amino acids linked by peptide bonds; a breakdown or build-up unit in protein metabolism.

Peptide bond A chemical bond holding amino acid sub-units together in proteins.

Peptide vaccine A short chain of amino acids that can induce antibodies against a specific infectious agent.

Peptidyl-tRNA binding site; P-site The site on a ribosome that hosts the tRNA to which an amino acid for the growing polypeptide chain is attached.

Peptidyl transferase An enzyme bound tightly to the large sub-unit of the ribosome that catalyses the formation of peptide bonds between amino acids during translation.

Perennial (L. *perennis*, lasting years) A plant that grows more or less indefinitely from year to year and, once mature, usually produces seed each year.

Phytochrome A reversible pigment system of a protein nature, found in the cytoplasm of green plants. Phytochrome is associated with the absorption of light that affects growth, development and differentiation of a plant, independent of photosynthesis, e.g., in the photoperiodic response.

Phytohormone A substance that stimulates growth or other processes in plants. They include auxins, abscissic acid, cytokinins, gibberellins and ethylene. Phytohormones are chemical messengers that may pass through cells, tissues and organs and stimulate biochemical, physiological and morphological responses.

Plant cell culture Growth of plant cells or roots of plants in bioreactors.

Polygenic Controlled by many genes of small effect.

Polylinker A segment of DNA that contains a number of different restriction endonuclease sites. a.k.a. multiple cloning site (MCS).

Polymer A compound composed of many identical smaller sub-units; resulting from the process of polymerization.

Polymerization Chemical union of two or more molecules of the same kind such as glucose or nucleotides to form a new compound (starch or nucleic acid) having the same elements in the same proportions but a higher molecular weight and different physical properties.

Polymery The phenomenon whereby a number of genes at different loci (which may be polygenes) can act together to produce a single effect.

Polymorphism The occurrence of two or more alleles at a locus in a population. a.k.a. genetic polymorphism.

Polysaccharide (Gr. *polys*, many + Gr. *sakcharon*, sugar) Long-chain molecules, such as starch and cellulose, composed of multiple units of a monosaccharide.

Polysaccharide capsules Carbohydrate coverings with antigenic specificity that are present on some types of bacteria.

Polytene chromosomes Giant chromosomes produced by interphase replication without division, and consisting of many identical chromatids arranged side by side in a cablelike pattern.

Polyunsaturates Oils containing polyunsaturated fatty acids.

Polyspermy The entry of several sperm into the egg during fertilization, although only one sperm nucleus actually fuses with the egg nucleus.

Preventive immunization; vaccination Infection with an antigen to elicit an antibody response that will protect the organism against future infections.

Prolactin A hormone, produced by the anterior pituitary gland, that stimulates and controls lactation in mammals.

Proliferation (L. *proles*, offspring + *ferre*, to bear) Increase by frequent and repeated reproduction; growth by cell division.

Pro-meristem The embryonic meristem that is the source of organ initials or foundation cells.

Pro-nucleus Either of the two haploid gamete nuclei just prior to their fusion in the fertilized ovum.

Proofreading The enzymatic scanning of newly-synthesized DNA for structural defects, such as mis-matched base pairs. It is effected by DNA polymerase.

Propagation (L. *propagare*, to propagate) The multiplication of plants by numerous types of vegetative material; an ancient practice dating from the dawn of agriculture, carried out in a nursery or directly in the field (vegetative propagation), and now in *in vitro* culture (micropropagation).

Propagule Any structure capable of giving rise to a new plant by asexual or sexual reproduction, including bulbils, leafbuds, etc.

Pro-phage The genome of a temperate bacteriophage integrated into the chromosome of a lysogenic bacterium and replicated along with the host chromosome.

Prophase (Gr. *pro*, before + *phasis*, appearance) An early stage in nuclear division, characterized by the shortening and thickening of the chromosomes and their movement to the metaphase plate. It occurs between interphase and metaphase. During this phase, the centriole divides and the two daughter centrioles move apart. Each sister DNA strand from interphase replication becomes coiled, and the chromosome is longitudinally double except in the region of the centromere. Each partially separated chromosome is called a chromatid. The two chromatids of a chromosome are sister chromatids.

Protamines Small basic proteins that replace the histones in the chromosomes of some sperm cells.

Protease An enzyme that hydrolyzes proteins, cleaving the peptide bonds that link amino acids in protein molecules.

Protein crystallization Making crystals of a protein, as a key part of most methods used for determining a protein's three-dimensional structure.

Protein engineering Generating proteins with modified structures that confer properties such as higher catalytic specificity or thermal stability.

Protein kinase An enzyme that adds phosphate groups to a protein molecule at the sites of serine, threonine or tyrosine residues.

Protein metabolic step One step in the chain of reactions that take place in an organism and dictate the composition of that organism.

Protein sequencing The process of determining the amino-acid sequence of a protein or its component polypeptides.

Protein synthesis The creation of proteins from their constituent amino acids, in accordance with the genetic information carried in the DNA of the chromosomes.

Proteinaceous infectious particle; prion An abnormal form of a normal cell protein, with no detectable nucleic acid, found in the brain of mammals and believed to be the agent responsible for the class of diseases called spongiform encephalopathies, including scrapie in sheep and bovine spongiform encephalopathy (BSE; mad cow disease) in cattle.

Proteolysis Enzymatic degradation of a protein.

Proteolytic The ability to break down protein molecules.

Protoclone Regenerated plant derived from protoplast culture or a single colony derived from protoplasts in culture.

Protocol (Gr. *protokollon*, first leaf glued to a manuscript and describing the contents) The step-by-step experiments proposed to describe or solve a scientific problem, or the defined steps of a specific procedure.

Protocorm In orchids, seeds contain an unorganized embryo comprising only a few hundred cells. During seed germination this embryo forms a tuberous structure, called a protocorm, from which develops a complete plant. In cultures, vegetative parts of several orchids form round, smooth protocorms which may be multiplied indefinitely or induced to regenerate a whole plant.

Protoderm (Gr. *protos*, first + *derma*, skin) A primary meristem tissue that gives rise to epidermis.

Protogyny The condition in which the female reproductive organs (carpels) of a flower mature before the male ones (stamens), thereby ensuring that self-fertilization does not occur.

Protoplasm The essential, complex living substance of cells, upon which all vital functions of nutrition, secretion, growth and reproduction depend.

Protoplast (Gr. *protoplastos*, formed first) A bacterial or plant cell for which the relatively rigid wall has been removed either chemically or enzymatically, leaving its cytoplasm enveloped by only a delicate peripheral membrane. Protoplasts are spherical and smaller than the elongate, angular shaped and often vacuolated cells from which they have been released.

Protoplast culture The isolation and culture of plant protoplasts by mechanical means or by enzymatic digestion of plant tissues or organs, or cultures derived from these. Protoplasts are utilized for selection or hybridization at the cellular level and for a variety of other purposes.

Protoplast fusion The coalescence of the plasmalemma and cytoplasm of two or more protoplasts in contact with one another. Initial adhesion of two protoplasts is a random process but coalescence can be induced.

Prototroph An organism such as a bacterium that will grow on a minimal medium.

Protozoan (pl: protozoa) A microscopic, single-cell organism.

Pulsed-field gel electrophoresis (PFGE) A procedure used to separate very large DNA molecules by alternating the direction of electric current in a pulsed manner across a semisolid gel.

Punctuated equilibrium The occurrence of speciation events in bursts, separated by long intervals of species stability.

Pure culture Axenic culture.

Pure line A strain in which all members have descended by self-fertilization or close inbreeding . A pure line is genetically uniform.

Purine A double-ring, nitrogen-containing base present in nucleic acids; adenine (A) and guanine (G) are the two purines present in most DNA and RNA molecules.

Quantitative genetics The area of genetics concerned with the inheritance of continuously-varying traits. Most practical improvement programs involve the application of quantitative genetics.

Quantitative inheritance Inheritance of measurable traits (height, weight, colour intensity, etc.) that depend on the cumulative action of many genes.

Quantitative trait A measurable trait that shows continuous variation; a trait that can not be classified into a few discrete classes.

Quantitative trait locus (QTL) A locus that affects a quantitative trait. The plural form (quantitative trait loci) is also abbreviated as QTL.

Quantum (L. *quantum*, how much) An elemental unit of energy. Its energy value is hv, where h, Planck's constant, is 6.62×10^{-27} erg-second and v is the frequency of the vibrations or waves with which the energy is associated. *cf* photon.

Quantum speciation The rapid formation of new species, primarily by genetic drift (q.v.).

Quarantine (It. *quarantina*, from *quaranta,* forty) Originally, keeping a person or living organism in isolation for a period (originally 40 days) after arrival to allow disease symptoms to appear, if there was any disease present. Now used for regulations restricting the sale or shipment of living organisms, usually to prevent disease or pest invasion of an area.

Quiescent Quiet, at rest, but not necessarily dormant, and having the potential for resumed activity; can apply to non-meristematic cells.

Raceme An inflorescence in which the main axis is elongated but the flowers are borne on pedicels that are about equal in length.

Rachilla (Gr. *rhachis*, a backbone + L. diminutive suffix *-illa*) Shortened axis of a spikelet.

Rachis (Gr. *rhachis*, a backbone) Main axis of a spike; axis of fern leaf (frond) from which pinnae arise; in compound leaves, the extension of the petiole corresponding to the midrib of an entire leaf.

Radicle (L. *radix*, root) That portion of the plant embryo which develops into the primary or seed root.

Radioactive isotope; radioisotope An unstable isotope that emits ionizing radiation.

Ramet An individual member of a clone. *cf* ortet.

Random amplified polymorphic DNA (RAPD; pronounced 'rapid') A technique using single, short (usually 10-mer) synthetic oligonucleotide primers for PCR. The primer, whose sequence has been chosen at random, initiates replication at its complementary sites on the DNA, producing fragments up to about 2 kb long, which can be separated by electrophoresis and stained with ethidium bromide. A primer can exhibit polymorphism between individuals, and polymorphic fragments can be used as markers.

Random mutagenesis A non-directed change of one or more nucleotide pairs in a DNA molecule.

Receptacle (L. *receptaculum*, a reservoir) Enlarged end of the pedicel or peduncle, to which other flower parts are attached.

Receptor A molecule that can accept the binding of a ligand.

Receptor-binding screening One of the biotechnology-based methods for discovering conventional drugs. The method relies on the fact that many drugs act by binding to specific proteins (receptors) on or in cells: these proteins usually bind to hormones or to other cells, and control the cell's behaviour, although they may be enzymes or structural elements of the cell. The drug interferes with the normal role of the protein.

Recessive Describing an allele whose effect with respect to a particular trait is not evident in heterozygotes. Opposite to *dominant*.

Reciprocal crosses Crosses between the same two strains, but with the sexes reversed; e.g., female A × male B and male A × female B.

Reciprocating shaker A platform shaker used for agitating culture flasks, with a back and forth action at variable speeds.

Recognition site A nucleotide sequence - composed typically of 4, 6 or 8 nucleotides - that is recognized by and to which a restriction endonuclease (restriction enzyme) binds. For type II restriction enzymes (those used in gene-cloning experiments) it is also the sequence within which the enzyme specifically cuts (and their corresponding enzymes methylate) the DNA, i.e., for type II enzymes, the recognition site and the target site are the same sequence. Type I enzymes bind to their recognition site and then cleave the DNA at some more or less random position outside that recognition site. *cf* restriction site.

Reporter gene A gene that encodes a product that can readily be assayed. Thus reporter genes are used to determinate whether a particular DNA construct has been successfully introduced into a cell, organ or tissue.

Repressible enzyme An enzyme whose synthesis is diminished by a regulatory molecule.

Saccharifaction Following liquefaction, the hydrolysis of polysaccharides by glucoamylase to maltose and glucose.

Salmonella A genus of rod-shaped, Gram-negative bacteria that are a common cause of food poisoning.

Salt tolerance; saline resistance The ability to withstand a concentration of sodium (Na^+ ion), or of any other salt, in the soil (or in culture), which is damaging or lethal to other plants. Breeding and selection for increased tolerance and resistance in crop plants is of great current interest.

Sap Fluid content of the xylem and phloem cells of plants. Fluid contents of the vacuole are referred to as cell sap.

Saprophyte A vegetable organism that derives its nutriment from decaying organic matter.

Satellite DNA That portion of the DNA in plant and animal cells consisting of highly repetitious sequences (millions of copies) typically in the range from 5 to 500 bases. Thousands of copies occur tandemly (end-on-end) at each of many sites. It can be isolated from the rest of the DNA by density gradient centrifugation.

Satellite RNA A small, self-splicing RNA molecule that accompanies several plant viruses, including tobacco ringspot virus. a.k.a. viroid.

Saturates Noun: Oils containing saturated fatty acids.

Scaffold The central core structure of condensed chromosomes. The scaffold is composed of non-histone chromosomal proteins.

Scale up Conversion of a process, such as fermentation of a micro-organism, from a small scale to a larger scale.

Scanning electron microscope (SEM) An electron-beam-based microscope used to examine, in a three dimensional screen image, the surface structure of prepared specimens.

Scarification The chemical or physical treatment given to some seeds (where the seed coats are very hard or contain germination inhibitors) in order to break or weaken the seed coat sufficiently to permit germination.

Scientific name A unique identifier consisting of a genus and a species name (the specific epithet) in Latin, assigned to each recognized and described species of organism. Based on the Linnean system of classification. Susceptible to considerable blurring at the edges because, as Darwin came to realize, nature has not packaged living things into neatly discrete entities. *cf* cultivar; pathovar.

Scion The twig or bud to be grafted onto another plant, the root stock, in a budding or grafting operation.

Scion-stock interaction The effect of a rootstock on a scion (and vice versa) in which a scion on one kind of rootstock performs differently than it would on its own roots or on a different rootstock.

Target site duplication A sequence of DNA that is duplicated when a transposable element inserts; usually found at each end the insertion.

Targeted drug delivery A method of delivering a drug to the site in the body where it is needed, rather than allowing it to diffuse into many sites.

Targeting vector A cloning vector carrying a DNA sequence capable of participating in a crossing-over event at a specified chromosomal location in the host cell.

Tautomeric shift The transfer of a hydrogen atom from one position in an organic molecule to another position.

Tautomerism A type of isomerism in which the two isomes are in equilibrium.

T cells; T lymphocyte Lymphocyte that pass through the thymus gland during maturation. Different kinds of T cells play important roles in the immune response, being primarily responsible for the T cell-mediated response or cellular immune response.

T cell receptor An antigen-binding protein that is located on the surfaces of killer T cells and mediates the cellular immune response of mammals. The genes that encode T cell antigens are assembled from gene segments by somatic recombination processes that occur during T lymphocyte differentiation.

T-cell-mediated (cellular) immune response The synthesis of antigen-specific T cell receptors and the development of killer T cells in response to an encounter of immune system cells with a foreign immunogen.

T-DNA The segment of DNA in the Ti plasmid of *Agrobacterium tumefaciens* that is transferred to plant cells and inserted into the chromosomes of the plant.

T4 DNA ligase An enzyme from bacteriophage-T4-infected cells, and that catalyses the joining of duplex DNA molecules and repairs nicks in DNA molecules. The enzyme requires that one of the DNA molecules has a 5´-phosphate group and that the other has a free 3´-hydroxyl group.

TRNA; transfer RNA The class of small RNA molecules that transfer amino acids to the ribosome during protein synthesis. Transfer RNA molecules are folded into a 'clover-leaf' secondary structure by intrastrand base pairing. The anticodon loop contains a nucleotide-triplet complementary to a specific codon within the mRNA molecule. Each tRNA is 'charged' with the correct amino acid molecule, via its 3´ adenosine moiety, by an enzyme called aminoacyl-tRNA synthetase.

Telomerase An enzyme that adds telomeric sequences to the ends of eukaryotic chromosomes.

Telemeter The unique structure found at the end of eukaryotic chromosomes containing specialized sequences of DNA that assures the completion of a cycle of DNA replication.

Telophase (Gr. *telos*, end + phase) The last stage in each mitotic or meiotic division, in which the chromosomes are assembled at the poles of the division spindle.

Ultraviolet light; ultraviolet radiation (UV) The portion of the electromagnetic spectrum with wavelengths from about 100 to 400 nm; between ionizing radiation (X-rays) and visible light. UV is absorbed by DNA and is highly mutagenic to unicellular organisms and to the epidermal cells of multicellular organisms. UV light is used in tissue culture for its mutagenic and bactericidal properties.

Undefined A medium or substance added to medium in which not all of the constituents or their concentrations are chemically defined, such as media containing coconut milk, malt extract, casein hydrolysate, fish emulsion or other complex compounds. *cf* organic complex.

Understock Host plant for a grafted scion, a branch or shoot from another plant; an understock may be a fully grown tree or a stump with a living root system.

Undifferentiated In a meristematic state or resembling a meristem; lacking the specialized or differential gene expression characteristic of specialized cells.

Unequal crossing over Crossing over between repeated DNA sequences that have paired out of register, creating duplicated and deficient products.

Unicellular Describing tissues, organs or organisms consisting of a single cell.

Unisexual Describing animals and plants possessing either male or female reproductive organs, but not both.

Univalent An unpaired chromosome at meiosis.

Universal donor cells Cells that, after introduction into a recipient, will not induce an immune response that leads to their rejection.

Universality Referring to the genetic code, the codons have the same meaning, with minor exceptions, in virtually all species.

Unorganized growth *In vitro* formation of tissues with few differentiated cell types and no recognizable structure; many call are unorganized. *cf* organized growth.

Vaccine A preparation of dead or weakened pathogens, or of derived antigenic determinants, that is used to induce formation of antibodies or immunity against the pathogen.

Vaccinia The cowpox virus used to vaccinate against smallpox and, experimentally, as a carrier of genes for antigenic determinants cloned from other disease organisms.

Vacuole (L. diminutive of *vacuus*, empty) A cavity in a plant cell, bounded by a membrane; in which various plant products and by-products are stored.

Vacuum Vacuum is created through use of a vacuum pump or aspirator pump, to facilitate specific biological preparations, such as inclusions or disinfection of material for *in vitro* culture, etc.

Variable domains Regions of antibody chains that have different amino acid sequences in different antibody molecules. These regions are responsible for the antigen-binding specificity of the antibody molecule.

Variable expressivity Variation in the phenotype caused by different alleles of the same gene and/or by the action of other genes and/or by the action of non-genetic factors. *cf* expressivity.

Variable surface glycoprotein (VSG) One of a battery of antigenic determinants expressed by a micro-organism to elude immune detection.

Variance In statistics: The sum of the squared deviations, divided by one less than the number of observations. A statistical measure of variation in a population.

Variant An organism that is genetically different from the wild type organism. a.k.a. mutant.

Variation Differences between individuals within a population or among populations.

Variegated Plants having both green and albino tissues. This difference in colour may result from viral infection, nutritional deficiency, or may be under genetic or physiological control.

Variety A naturally occurring subdivision of a species, with distinct morphological characters and given a Latin name according to the rules of the International Code of Nomenclature. A taxonomic variety is known by the first validly published name applied to it so that nomenclature tends to be stable. *cf* cultivar; pathovar.

Vitrified; water soaked Cultured tissue having leaves and sometimes stems with a glassy, transparent or wet and often swollen appearance. The process of vitrification is a general term for a variety of physiological disorders that lead to shoot tip and leaf necrosis.

Walking A method for cloning large regions of a chromosome. Starting from a known site, a gene library is screened for clones that hybridize to DNA probes taken from the ends of the first clone. These clones are then isolated, and their ends used to screen the library again. These clones are then isolated and their ends used, and so on.

Wet weight The gross weight of a product with its full water content, or weight of fully hydrated tissue. *See* fresh weight.

Wild type An organism as found in nature; the dominant allele usually found in nature, and from which mutations produce other dominants or recessives alleles.

Wilt Drooping of stems and foliage due to loss of water and decreased turgidity of cells. May be caused by water stress or by disease.

Wilting point (WP) The moisture content of soil at which plants start to wilt, but not to the extent that they fail to recover when placed in a humid atmosphere. *cf.* permanent wilting point.

Wobble hypothesis An explanation of how one tRNA may recognize more than one codon. The first two bases of the mRNA codon and anticodon pair properly, but the third base in the anticodon has some wobble that permits it to pair with more than one base.

Xanthophyll (Gr. *xanthos*, yellowish brown + *phyllon*, leaf) A yellow chloroplast pigment.

X-chromosome A chromosome associated with sex determination. In most animals, the female has two, and the male has one X chromosome. The opposite occurs in birds, in which the equivalent is the Z chromosome.

Xenia The immediate effect of pollen on some characters of the endosperm.

Xenobiotic A chemical compound that is not produced by living organisms; a manufactured chemical compound.

Xenotransplantation The transplantation of tissue from one species to another species, typically from non-human mammals to humans. This technology has become very important because of a worldwide shortage of human organs for humans requiring a new organ. The most popular non-human species involved in this technology is the pig.

Xerophyte (Gr. *xeros*, dry + *phyton*, a plant) A plant very resistant to drought or that lives in very dry places.

X-linked The presence of a gene on the X chromosome. a.k.a. X-linkage.

X-linked disease A genetic disease caused by an allele at a locus on the X-chromosome. In X-linked recessive conditions, a normal female "carrier" passes on the allele on her X chromosome to an affected son.

X-ray crystallography The deduction of crystal structure from analysis of the diffraction pattern of X-rays passing through a pure crystal of a substance.

Xylem (Gr. *xylon*, wood) A complex tissue specialized for efficient conduction of water and mineral nutrients in solution. Xylem may also function as a supporting tissue, particularly secondary xylem.

Y-chromosome The partner of the X-chromosome in the male of many animal species.

YAC Yeast artificial chromosome, used as a vector system for cloning DNA fragments that can be hundreds of kilobases long. Linear cloning vectors constructed from essential elements of yeast chromosomes. They can accommodate foreign DNA inserts of 200 to 500 kb.

Yeast episomal vector (YEp) A cloning vector for the yeast *Saccaromyces cerevisiae* that uses 2mm plasmid as origin of replication and is maintained as an extrachromosomal nuclear DNA molecule.

Yeast extract A mixture of substances from yeast. *cf* organic complex; undefined.

Yeast A unicellular ascomycete fungus, commonly found as a contaminant in plant tissue culture.

Yeast cloning vectors The yeasts, and especially *Saccaromyces cerevisiae*, are favourite organisms in which to clone and express DNA. They are eukaryotes, and so can splice out introns, the non-coding sequences in the middle of many eukaryotic genes.

Zone of elongation The section of the young root or shoot just behind the apical meristem, in which the cells are enlarging and elongating rapidly.

Zoo blot Hybridization of cloned DNA from one species to DNA from other organisms to determine the extent to which the cloned DNA is evolutionarily conserved.

Zoo FISH Fluorescent *in situ* hybridization of DNA from one species on metaphase chromosomes of another species. Typically, the hybridization is done separately for DNA libraries representing each chromosome. The result is a fascinating picture of the regions of chromosomal homology between species.

Zoospore A spore that possesses flagella and is therefore motile.

Zygonema (adj: zygotene) Stage in meiosis during which synapsis occurs; coming after the leptotene stage and before the pachytene stage in the meiotic prophase.

Zygospore (Gr. *zygon*, a yoke + spore) A thick-walled resistant spore developing from a zygote resulting from the fusion of isogametes.

Zygote (Gr. *zygon*, a yoke) A diploid cell formed by the fusion of two haploid gametes during fertilization in eukaryotic organisms with sexual reproduction. It is the first cell of the new individual.

Zymogen Inactive enzyme precursor that after secretion is chemically altered to the active form of the enzyme.

Bibliography

Ahmadian, A., A. Russom, H. Andersson, M. Uhlen, G. Stemme, and P. Nilsson (2002). SNP analysis by allele-specific extension in a micromachined filter chamber. *Biotechniques*, 32(4): p. 748, 750, 752, 754.

Alberts, Bruce; Johnson, A, Lewis, J, Raff, M, Roberts, K & Walter, P (2002). *Molecular Biology of the Cell*, 4th edition, Garland.

Andersson AC, Stromberg S, Backvall H, Kampf C, Uhlen M, Wester K, Ponten F. (2006). Analysis of protein expression in cell microarrays: a tool for antibody-based proteomics. *J Histochem Cytochem*. 54(12):1413-23.

Andrade, Berglund, Uhlen and Odeberg (2006). Using Grid Technology for Computationally Intensive Applied Bioinformatics Analyses, *In Silico Biology*, 495 - 504.

Bandmann, N., E. Collet, J. Leijen, M. Uhlen, A. Veide, and P.A. Nygren (2000). Genetic engineering of the Fusarium solani pisi lipase cutinase for enhanced partitioning in PEG-phosphate aqueous two-phase systems. *J Biotechnol*, 79(2): p. 161-72.

Begon, Michael; Townsend, CR & Harper, JL (2005). *Ecology: From Individuals to Ecosystems*, 4th edition, Blackwell Publishing Limited.

Bengtsson S, Krogh M, Szigyarto CA, Uhlen M, Schedvins K, Silfversward C, Linder S, Auer G, Alaiya A, James P. (2007). Large-Scale Proteomics Analysis of Human Ovarian Cancer for Biomarkers. *J Proteome Res*. 6(4):1440-1450.

Bledsoe, J. Gary (2000). "Ethical discussions in an undergraduate tissue engineering course".

Blomstergren, A., D. O'Meara, M. Lukacs, M. Uhlen, and J. Lundeberg (2000). Cooperative oligonucleotides in purification of cycle sequencing products. *Biotechniques*, 29(2): p. 352-4, 356, 358 passim.

Brock, T. and Madigan, M (1992). "Gene manipulation and genetic engineering," In: Biology of Microorganisms 6th Edition. Englewood Cliffs, NJ: Prentice Hall, pp. 280-305.

Bud, R. (1989). "Janus-faced biotechnology: An historical perspective." Trends in Biotechnology. 7: 230-233.

Campbell, Neil (2004). *Biology*, 7th edition, Benjamin-Cummings Publishing Company.

Cook, S.B. (2006). *Quantification of injury to aquatic resources resulting from the Pryor oil spill and fire, Obed Wild and Scenic River, Task 1* (pdf file). Final Draft Report.

Cook, S.B., Russ, W.T., and Goodfred, D.W. (2005). *Emory River Watershed biological assessment* (pdf file). Quarterly Report submitted to the U.S. Fish and Wildlife Service and Tennessee Wildlife Resources Agency.

Cook, S.B., Russ, W.T., and Goodfred, D.W. (2006). *Biological survey of the Emory River Watershed, Tennessee* (pdf file). Final Report submitted to the U.S. Fish and Wildlife Service and Tennessee Wildlife Resources Agency.

Cram, Nicholas, John Wheeler, and Charles S. Lessard (1995). "Ethical Issues of Life-sustaining Technology". *IEEE Technology and Society Magazine* 14.1, 21-28.

Fielder, J.H. (1992). "The Bioengineer's Obligations to Patients". *Journal of Investigative Surgery* 5.3, 201-8.

Grundfest, Warren S. and Andrea K. Scott (1998). "Ethical Issues in the Development of Medical Devices: The Role of Education and Training in the Evaluation of Safety and Efficacy". *Critical Reviews in Biomedical Engineering* 26.5-6, 378.

Hertzberg, M., H. Aspeborg, J. Schrader, A. Andersson, R. Erlandsson, K. Blomqvist, R. Bhalerao, M. Uhlen, T.T. Teeri, J. Lundeberg, B. Sundberg, P. Nilsson, and G. Sandberg (2001). A transcriptional roadmap to wood formation. *Proc Natl Acad Sci U S A*, 98(25): p. 14732-7.

Hjalmarsson S, Alderborn A, Fock C, Muldin I, Kling H, Uhlen M, Engstrand L. (2004). Rapid combined characterization of microorganism and host genotypes using a single technology. *Helicobacter*. 9(2):138-45.

Johansson T, Le Quere A, Ahren D, Soderstrom B, Erlandsson R, Lundeberg J, Uhlen M, Tunlid A.(2004). Transcriptional responses of Paxillus involutus and Betula pendula during formation of ectomycorrhizal root tissue. *Mol Plant Microbe Interact*. Feb;17(2):202-15.

Larsson, M., S. Stahl, M. Uhlen, and A. Wennborg (2000). Expression profile viewer (ExProView): a software tool for transcriptome analysis. *Genomics*, 63(3): p. 341-53.

Lerner M, Corcoran M, Cepeda D, Nielsen ML, Zubarev R, Ponten F, Uhlen M, Hober S, Grander D, Sangfelt O. (2007). The RBCC Gene RFP2 (Leu5) Encodes a Novel Transmembrane E3 Ubiquitin Ligase Involved in ERAD. *Mol Biol Cell*. 2007 Feb 21; [Epub ahead of print]

Levitan, M. (1988). Textbook of Human Genetics 3rd Edition. New York: Oxford Press.

Luria, S.E. and Delbruck, M. (1943) "Mutations of Bacteria from Virus Sensitivity to Virus Resistance." Genetics. 28:491-511. Mangino, R.A. (1975) A Laboratory Manual for Microbiology. Portland, ME: J. Weston Walch.

Lynn, Richard; Van Court, Marilyn (2004). "New evidence of dysgenic fertility for intelligence in the United States". *Intelligence* 32 (2): p. 193. Ablex Pub.

Magner, LN (2002). *A History of the Life Sciences*. TF-CRC.

Nilsson, P., A. Larsson, J. Lundeberg, M. Uhlen, and P.A. Nygren (1999). Mutational scanning of PCR products by subtractive oligonucleotide hybridization analysis. *Biotechniques*, 26(2): p. 308-16.

Nilsson, P., D. O'Meara, F. Edebratt, B. Persson, M. Uhlen, J. Lundeberg, and P. Nygren (1999). Quantitative investigation of the modular primer effect for DNA and peptide nucleic acid hexamers. *Anal Biochem*, 269(1): p. 155-61.

Parsons, J.W., and Benson, N.G. (1960). Fertilization of Obed River, Tennessee . *Journal of the Tennessee Academy of Science*, 3(1), 63-75.

Persson A, Hober S, Uhlen M. (2006). A human protein atlas based on antibody proteomics. *Curr Opin Mol Ther*. 8(3):185-90.

Ronnmark, J., H. Gronlund, M. Uhlen, and P.A. Nygren (2002). Human immunoglobulin A (IgA)-specific ligands from combinatorial engineering of protein A. *Eur J Biochem*, 269(11): p. 2647-55.

Speir, H.J. (1969). *Comparison of rock bass,* Ambloplites rupestris (Rafinesque) *populations in Obed River and Spring Creek, Tennessee*. Master's Thesis. Tennessee Technological University: Cookeville.

Stedman, S., and Stedman, B. (2002). *Notes on the birds of the Big South Fork National River and Recreation Area and the Obed River National Wild and Scenic River* . Cookeville: Tennessee Technological University.

Towle, Albert (1988). *Modern Biology*. Austin, TX: Holt, Rinehart and Winston.

Vener, T., A. Holmberg, J. Lundeberg, and M. Uhlen (1999). Selection analysis of single-stranded clones using solid-phase techniques. *Genet Anal,* 15(2): p. 47-50.

Vener, T., M. Nygren, A. Andersson, M. Uhlen, J. Albert, and J. Lundeberg (1998). Use of multiple competitors for quantification of human immunodeficiency virus type 1 RNA in plasma. *J Clin Microbiol*, 36(7): p. 1864-70.

Index